Marine Geology and Geotechnology of the South China Sea and Taiwan Strait

Ronald C. Chaney

CRC Press
Taylor & Francis Group
Boca Raton London New York

CRC Press is an imprint of the
Taylor & Francis Group, an **informa** business

First edition published 2021
by CRC Press
6000 Broken Sound Parkway NW, Suite 300, Boca Raton, FL 33487-2742

and by CRC Press
2 Park Square, Milton Park, Abingdon, Oxon, OX14 4RN

CRC Press is an imprint of Taylor & Francis Group, an Informa business

---

Library of Congress Cataloging-in-Publication Data
Names: Chaney, Ronald C., author.
Title: Marine geology and geotechnology of the South China Sea and Taiwan Strait / Ronald C. Chaney, Emeritus Professor of Environmental Resources Engineering, Humboldt State University, California.
Description: Boca Raton : CRC Press, 2021. | Includes index.
Identifiers: LCCN 2020026405 (print) | LCCN 2020026406 (ebook) | ISBN 9780367608729 (hardback) | ISBN 9781003102328 (ebook)
Subjects: LCSH: Marine ecology--South China Sea. | Marine ecology--Taiwan Strait. | Geotechnical engineering--South China Sea. | Geotechnical engineering--Taiwan Strait. | Oceanography--South China Sea. | Oceanography--Taiwan Strait.
Classification: LCC QH95.22 .C43 2021 (print) | LCC QH95.22 (ebook) | DDC 577.7--dc23
LC record available at https://lccn.loc.gov/2020026405
LC ebook record available at https://lccn.loc.gov/2020026406

ISBN 13: 978-0-367-60872-9 (hbk)
ISBN 13: 978-0-367-60873-6 (pbk)
ISBN 13: 978-1-003-10232-8 (ebk)

Typeset in Times
by SPi Global, India

Dedication

This book was begun in 2000 as an invited lecture on the Taiwan Strait: Marine Geology, Geotechnology and Seismicity given at Nanjing University in the People's Republic of China. The book took longer to write than most because of the usual issues: illness, retirement, other projects, and life. It is therefore respectively dedicated to the following individuals who in many ways influenced and helped with the writing of this book.

To my wife
Patricia Jane Chaney

and to
Prof. Hsai-Yang Fang
and
Prof. Kenneth L. Lee

Contents

Preface....xiii
Acknowledgments....xv
Author....xvii

PART I Politics and Resources

Chapter 1 Introduction: Geography, History, and Politics....3

1.1 Geography....3
1.2 Political Divisions and Territorial Claims....4
1.3 Resources....5
References....8

PART II Tectonics/Geology

Chapter 2 Geology....11

2.1 Introduction....11
2.1.1 Sunda Plate....12
2.1.2 Yangtze Plate....12
2.1.3 Philippine Mobile Belt....13
2.2 South China Sea Geology....14
2.2.1 Topographic Features....14
2.2.2 Opening of the South China Sea....16
2.2.2.1 Indian Subcontinent and South-east Extrusion....18
2.2.2.2 Spratly Islands....20
2.2.2.3 Dangerous Grounds....20
2.2.2.4 Seismicity....21
2.2.2.5 Sediment Transport....21
2.3 Taiwan Strait Geology....23
2.3.1 Topographic Features....23
2.3.2 Taiwan Strait Tectonics....24
2.3.3 Seismicity....25
2.3.4 Sediment Transport....28
References....29

PART III Oceanographic Factors

Chapter 3 Oceanographic Factors 33

3.1 Introduction 33
3.2 Earth System Processes 33
3.3 Winds and Storms 36
3.3.1 Stationary Earth 36
3.3.2 Rotating Earth 36
3.3.3 South East Asia Wind Pattern 37
3.3.4 Global Wind Pattern 37
3.4 Tides 40
3.4.1 Origin 40
3.4.2 Diurnal and Semi-Diurnal 41
3.4.3 Tidal Currents 42
3.5 Currents 43
3.5.1 Surface Wind-Driven Currents 43
3.5.1.1 Wind-Driven Circulation 43
3.5.1.2 The Oceanic Gyres and Geostrophic Currents 46
3.5.1.3 Westward Intensification of Ocean Currents 47
3.5.1.4 Ekman Transport and Vertical Water Movement 47
3.5.1.5 Wind-Driven Circulation in the North Pacific 48
3.5.2 Vertical Currents, Density Driven 49
3.5.2.1 Thermalhaline Circulation 49
3.5.3 Water Exchange with Oceans 50
References 52

Chapter 4 Wave Characteristics 53

4.1 Introduction 53
4.2 Gravity Wave Generation 54
4.2.1 Linear Wave Theory (Airy Theory) 54
4.2.1.1 For Deep Water kh > π, $1/2 < h/\lambda < \infty$ 57
4.2.1.2 For Shallow Water, kh < $\pi/10$ or $h/\lambda < 1/20$ 57
4.2.2 Stokes Wave Theory 58
4.2.3 Cnoidal Wave Theory 60
4.2.3.1 Fetch and Seas 60
4.3 Wave Heights in Deep Water 61
4.4 Sea State 61
4.5 Earthquakes and Tsunamis 61

4.6 Storm Surges 64
4.6.1 Wind Fastest Mile 64
4.6.2 Gust Velocity 65
References 66

PART IV Terrestrial and Seabed Sediments

Chapter 5 Characterization of Marine Sediments 69

5.1 Introduction 69
5.2 Characterization of Sediments 71
5.2.1 Introduction 71
5.2.2 Clay-Like Material 72
5.2.3 Clay Microstructure 75
5.2.4 Texture, Structure, and Composition of Soil Particles 76
5.2.5 Clastic Soils 77
5.2.5.1 Size, Shapes, and Structure Disintegration 77
5.3 Classification of Sediments 79
5.3.1 Systems of Classification 79
5.3.1.1 Engineering Classification of Siliciclastic Sediments 81
5.3.1.2 Classification of Carbonate Sediments 82
References 88

Chapter 6 Index, Compressibility, and Strength of Marine Sediments 91

6.1 Introduction 91
6.2 Clays and Fine-Grained Sediments 92
6.2.1 Introduction 92
6.2.2 Density and Water Content 94
6.2.3 Consistency 94
6.2.4 Compressibility 96
6.2.5 Permeability 99
6.2.6 Shear Strength 105
6.3 Calcareous Materials 112
6.3.1 Introduction 112
6.3.2 Density and Water Content 112
6.3.2.1 Size Ranges of Particles 116
6.3.3 Consistency 116
6.3.3.1 Effect of Increasing Calcium Content 116
6.3.3.2 Water Content Changes Due to Particle Crushing 118
6.3.4 Compressibility 119

6.3.4.1 Primary Consolidation ... 119
6.3.4.2 Secondary Consolidation ... 123
6.3.5 Permeability ... 125
6.3.6 Shear Strength ... 126
6.3.7 Deformation Parameters ... 131
6.4 Siliceous Sediments ... 133
6.4.1 Introduction ... 133
6.4.2 Density and Water Content ... 136
6.4.3 Consistency ... 136
6.4.4 Compressibility ... 136
6.4.5 Permeability ... 136
6.4.6 Shear Strength ... 138
6.5 Organic Sediments ... 138
6.5.1 Introduction ... 138
6.5.1.1 Gel ... 139
6.5.1.2 Marine Clays and Organic Material in South East Asia ... 139
6.5.2 Density and Water Content ... 141
6.5.3 Consistency ... 142
6.5.4 Compressibility/Compaction ... 146
6.5.5 Permeability ... 151
6.5.6 Bioturbation ... 151
6.5.6.1 Introduction ... 151
6.5.6.2 Shallow Water, Benthos and Sediments ... 154
6.5.6.3 Bioturbated Deep Sea Sediments ... 154
References ... 155

Chapter 7 Behavior of Marine Sediments under Sustained, Dynamic and Cyclic Loading ... 161

7.1 Introduction ... 161
7.2 Sustained Loading ... 161
7.2.1 Viscoelastic Model Approach ... 164
7.2.2 Rate Process Approach ... 165
7.2.3 Empirical Approach ... 166
7.3 Dynamic and Cyclic Loading ... 168
7.3.1 Theoretical Background ... 168
7.3.2 Threshold Strain Concept ... 170
7.3.3 Clay Minerals under Cyclic Loading ... 173
7.3.4 Sands under Cyclic Loading ... 175
7.4 Soil Parameter Modeling ... 175
7.5 Behavior of Clays and Silts ... 181
7.5.1 Strength Determination ... 182
7.5.2 Pore Pressure Build-Up ... 183

7.5.3 Reduction and Degradation of Stiffness 187
7.5.3.1 Effect of Soil Plasticity on Cyclic Response...189
7.5.3.2 Factors That Affect Measurement of Dynamic Response of Clays 194
7.6 Analytical Methods to Predict Cyclic Response of Clays 198
7.7 Behavior of Granular Materials..198
7.7.1 Mechanism and Implications of Liquefaction Phenomena 200
7.7.2 Variation in Liquefaction 202
7.8 Wave Interaction with Seabed 203
7.8.1 Linear (Airy) Theory 203
7.8.2 Rigid Seabed..206
7.8.3 Deformable Seabed..207
7.9 Post-Cyclic Loading Behavior 209
References 210

Chapter 8 Sedimentation 217

8.1 Introduction 217
8.2 Shelf Sedimentation Patterns 217
8.3 Deep Basin Sedimentation 220
8.4 China Seas Sedimentation..220
8.5 Sediment Distribution Patterns in the China Seas...................221
References 221

PART V Civil Engineering Development

Chapter 9 Taiwan Strait Transportation (TST) Corridor 225

9.1 Introduction 225
9.2 Pacific Ocean Sea-floor Characteristics 225
9.3 Estimation of Geotechnical Parameters 226
9.4 Route Selection 226
9.5 Type of Construction...232
9.5.1 Tunnel 232
9.5.2 Bridge 233
9.5.2.1 Conceptional Design 233
9.5.2.2 Bridge Types 233
9.5.2.3 Aesthetics—Harmonizing with Surroundings 234
9.5.2.4. Loading 234
9.6 Sustainable Construction...238
References 238

Index...241

Preface

Sixty-five years have passed since the landmark paper on marine landslides was published by Karl von Terzaghi in 1956. Between 1956 and today, the world has seen significant changes in the expanding fields of marine geotechnology and geology and the use of the world's seabed. One-third of the world's shipping travels through the South China Sea and Taiwan Strait. This increased use of the shipping lane results in more exploitation of the continental shelf, slope, and abyssal plains of the China Seas. This usage has occurred as a result of increasing population, a desire to maintain or increase living standards, industrial and resource extraction activities, and security concerns. An understanding of the complicated marine environment is required to balance the cycles of energy and material exploitation along with security. The study of the seafloor environment and its relationship with engineering requires knowledge of other disciplines: soil science, geotechnical engineering, marine geology and geophysics, physical chemistry, mineralogy, and microbiology, to name just a few.

The primary purpose of this book is to provide a broad synthesis of concepts intended to serve students, teachers, and professional engineers and geologists on both marine geotechnology and geology of the South China Sea and Taiwan Strait. The author believes that a text encompassing the marginal seas of the Western Pacific needs to broadly cover the areas of geology, ocean driving mechanisms (i.e., tides, currents), wave mechanics, and physical properties of the various sediments that will be encountered. In addition, this text discusses the proposed Taiwan Strait crossing. This consists of various proposed design scenarios (i.e., combination of tunnels and bridges) and issues that should be addressed (i.e. aesthetics, wind, seismicity, vessel collision among a few).

Ronald C. Chaney
Emeritus Professor
Humboldt State University, California

Acknowledgments

Many individuals have helped put this book together in a variety of ways over the years. The following are a few of those individuals whom the author would like to thank: Dr. Kenneth R. Demars, Dr. Donald G. Anderson, Dr. H. Y. Fang, Dr. Armand J. Silva, Dr. Vincent P. Drnevich, Dr. Homa J. Lee, Dr. William R. Bryant, Dr. George H. Keller, Dr. Richard H. Bennett, Dr. Adrian Richards, Dr. Kathryn Moran, Dr. Umesh Dayal, Dr. Gideon Almagor, Dr. Geoff Martin, Prof. Shi Bin, Dr. Jian-Hua Yin, Dr. Bing Yen, Dr. Funan Peng, Dr. Zhiming Wu, Prof. Gao Guorui, Prof. Yuanbo Liang, Dr. Nitin Pandit, Richard S. Ladd, Willard L. DeGroff, Ronald J. Ebelhar, S.M. Slonim, and S.S. Slonim.

Author

Ronald C. Chaney, PhD, is emeritus professor of environmental resources engineering and former director of the Telonicher Marine Laboratory and head of vessel operations at Humboldt State University in Arcata, California. Previously, he was an associate professor of civil engineering and associate director of the marine geotechnical laboratory at Lehigh University in Bethlehem, Pennsylvania. He earned his PhD in engineering from the University of California, Los Angeles, in 1978. Dr. Chaney is a licensed civil and geotechnical engineer in California. He was the editor of the following journals: *International Journal of Marine Geotechnology*, *Journal of Marine Georesources and Geotechnology*, and the American Society for Testing and Materials *Geotechnical Testing Journal*. He is a fellow of both American Society of Civil Engineers (ASCE) and American Society for Testing and Materials (ASTM). He is also the co-author of the books *Seafloor Processes and Geotechnology* and *Introduction to Environmental Geotechnology 2nd edition*, both published by CRC Press.

Part I

Politics and Resources

1 Introduction: Geography, History, and Politics

1.1 GEOGRAPHY

The South China Sea (SCS) spans from Singapore and the Strait of Malacca in the south to the Taiwan Strait in the north, and from Borneo and the Philippines in the east to Vietnam and south-eastern China in the west (Figure 1.1). It is one of the world's largest marginal seas bordering eight countries. The SCS is important, among other things, for its international marine trust for transportation and navigation, its rich marine diversity (including fisheries), its impact on the

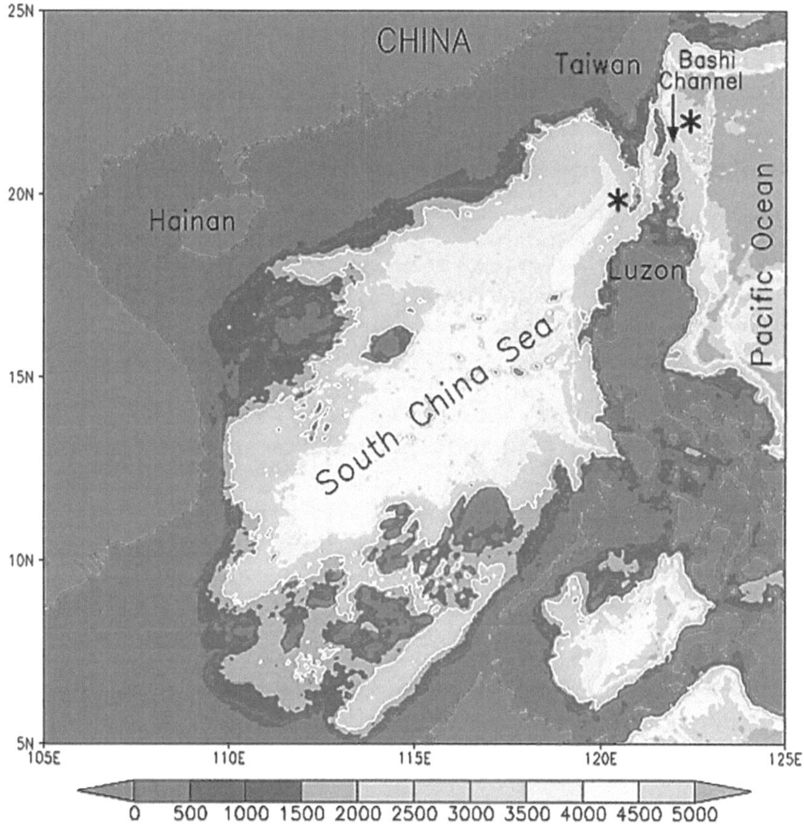

FIGURE 1.1 Schematic figure showing both the South China Sea and Taiwan Strait (Qu et al. 2006) Reprinted with permission of AGU.

monsoon climate of Southeast Asian countries, and its mineral resources. The SCS has numerous seamounts and islands. These are mainly coral reefs, atolls, shoals, and sand bars. Many of these features are submerged at high tides, and have no native inhabitants. They are usually clustered into geographical groups, notably the Spratly Islands and Reed Bank offshore of the Philippines, the Scarborough Seamount along the central axis of the SCS, and the Macclesfield Bank and the Paracel and Pratas Islands in the north. Collectively they have a total land area of less than 15 km^2 at low tide. The islands are located on a shallow continental shelf with an average depth of 200 m. There are exceptions in the Spratlys, the sea floor drastically changes its depth, and near the Philippines, the Palawan Trough is more than 5000 m deep. In addition, there are some parts that are so shallow that navigation becomes difficult.

1.2 POLITICAL DIVISIONS AND TERRITORIAL CLAIMS

Over the millennium, the SCS has served as both a fishery and a pathway for trade, exploration, and conquest. A timeline illustrating the political events affecting the SCS is presented in Table 1.1. At the present time, there are a number of competing

TABLE 1.1
Political Timeline of the SCS

- 1405–1433 Zheng He commanded seven expeditionary voyages from Ming dynasty China to Southeast Asia, South Asia, Western Asia, and East Africa. The first six voyages were authorized by the Yongle Emperor and the seventh during the Xuande reign.
- 1421 Zheng prepared the 6th edition of the Mao Kun map usually referred to as Zheng He's Navigation Map, which included the SCS islands.
- Nineteenth century France claimed control of the Spratlys until the 1930s as part of the occupation of Indochina. Several of these islands were exchanged with Britain.
- 1932–1935, the Republic of China (ROC) named 132 of the SCS islands.
- 1933, France occupied Taiping Island, ROC lodged formal protest.
- During WWII islands annexed by Japan.
- Japanese and the French renounced their claims as soon as their respective occupations or colonizations ended.
- 1947 the Republic of China (ROC) Ministry of Interior renamed 149 of the islands.
- 1947 November the Secretariat of Guangdong Government was authorized to publish a map of the SCS islands.
- 1949 People's Republic of China (PRC) claimed the islands as part of the province of Canton (Guangdong), which was changed later to the Hainan special administrative region.
- 1956 United Nations held a conference on the Law of the Sea. This conference resulted in a number of treaties signed in 1958, such as the Convention on the Territorial Sea and Contiguous Zone. The PRC was not a signatory because they were not a member of the UN at this time.
- 1958 PRC issued a declaration defining its territorial waters within what is known as the nine-dotted line which encompassed the Spratly Islands.
- 1958 Democratic Republic of Vietnam (DRV) sent a diplomatic note stating that they respected the PRC declaration. In this note, they indicated that they supported a 12-mile territorial sea along its territory.

(*Continued*)

TABLE 1.1 (*Continued*)
Political Timeline of the SCS

- 1959 PRC put Islands in the SCS under an administrative office.
- 1974 PRC seized Paracel Islands from Vietnam.
- 1988 PRC switched the administration office to the newly founded Hainan Province.
- 1988 Battle of Johnson Reef, China versus Philippines.
- 1988 PRC seized several of the Spratly Islands resulting in the sinking of Vietnamese ships killing at least 70 sailors.
- 2010 March, PRC told US officials that they considered the SCS a "core interest" on par with Taiwan, Tibet, and Xinjiang.
- 2010 July, PRC Communist Party-controlled *Global Times* newspaper stated that "China will never waive in its right to protect its core interest with military means" and the Ministry of Defense spokesman said that "China has indisputable sovereignty of the SCS and China has sufficient historical and legal backing" to underpin its claims.
- July 2012 *Half Moon Shoal*, Chinese naval frigate ran aground 60 miles off coast of Palawan province.
- Philippines submitted fisheries case to International Tribune for the Law of the Sea (ITLOS).
- 2013 China added a tenth-dash line to the east of Taiwan Island as a part of its official sovereignty claim to the disputed territories in the SCS.
- May 2014, 11 Chinese fishermen apprehended at Half Moon Bay.
- August 2014, China reclaiming land at Fiery Cross Reef. Area large enough for 3000 ft. airstrip.
- 2017 to present—various claims of sovereignty unresolved.

Source: Adapted from Wikipedia.

claims over whose territory, if any, the marginal sea belongs to. The sea areas recognized in international rights courts are shown in Figure 1.2. A review of this figure shows that territorial waters extend from a shoreline out 12 nautical miles. This is the distance that an eighteenth-century cannon can shoot. The shoreline is defined as the mean low water mark for the purposes of this discussion. The contiguous zone extends from the end of the territorial waters and an additional 12 nautical miles. The exclusive economic zone (EEZ) extends from the shoreline out 200 nautical miles. International waters in maritime law is defined as extending from the end of territorial water seaward. Typically, under maritime law in order to claim sovereignty over an area water the country making the claim must demonstrate exclusive control for a continuous period of time up to the present. No country at the present time bordering the SCS can demonstrate this requirement. A figure presenting the various claims is shown in Figure 1.3.

1.3 RESOURCES

The SCS was surrounded by some of the world's fastest developing nations until the Asian financial crisis in 1997. This semi-enclosed sea is a 1.4 million square miles body of water that carries roughly one-third of the world's shipping and could hold trillions of dollars in undersea deposits of oil and natural gas. The resources

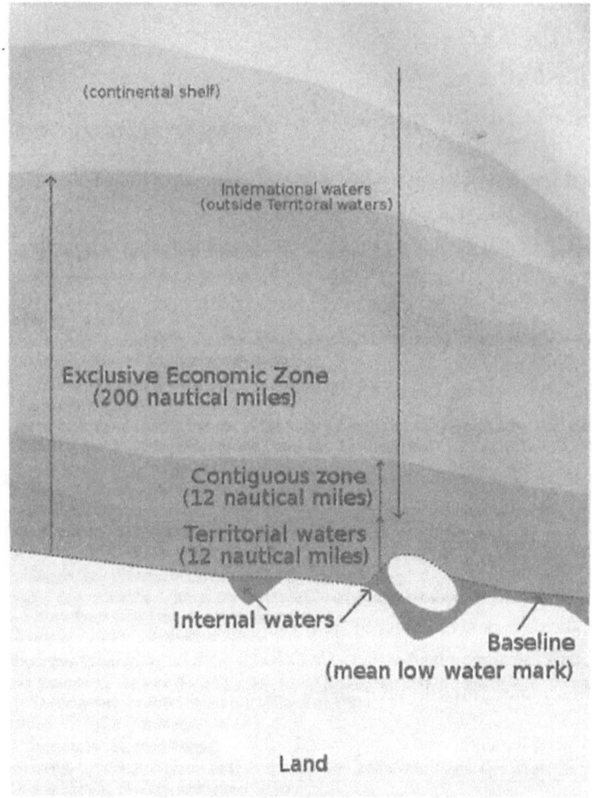

FIGURE 1.2 Sea areas recognized in international rights (Adapted from Image Zones maritimes UNCLS.jpg).

in this area are not equally distributed as shown in Figure 1.4 (Dahlby, 1998). These resources range from undersea deposits of oil fields, natural gas fields, and fisheries. Conflicting claims to these resources have resulted in open conflict. Overfishing has exhausted catches close to shorelines, and economic growth has outpaced existing oil supplies. The sovereignty over rocks, shoals, and reefs have been claimed by various countries to establish national outposts for asserting ownership of fishing grounds and the petroleum believed to lie beneath. An example of these disputes is the outright armed conflict between the People's Republic of China (PRC) and other countries bordering the SCS over offshore drilling rights along the Vietnamese and Philippine coasts. The real problem is that years of costly exploration have produced exasperatingly little oil. The Spratlys Islands are important because of their location, because they lie along one of the most strategic shipping routes in the world.

Under the United Nations convention for the Law of the Sea countries may designate areas within 200 nautical miles of their coasts over which a state has special

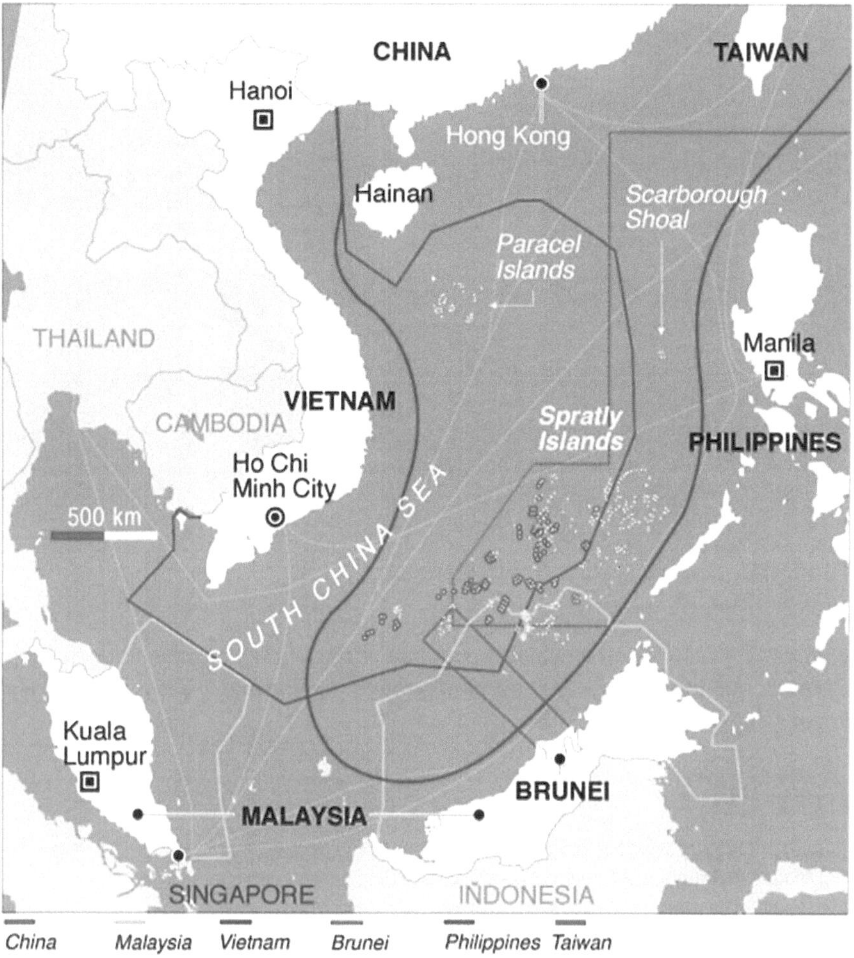

FIGURE 1.3 Competing countries' territorial claims in the South China Sea (Wikipedia Public Domain).

rights as exclusive economic zones (EEZ). Across the SCS, the EEZ zones overlap in some areas such as the Spratly Islands. Unfortunately, this has led to the EEZ claims asserted by the PRC overlapping a number of historical claims by other countries surrounding the SCS.

The area is poorly charted making it dangerous to navigate. It is called a dangerous ground partly due to the above navigation issues and because all the neighboring countries (Philippines, Vietnam, Malaysia, Brunei, China, and Taiwan) partly claim the area because of natural resources. According to current US estimates, the seabed beneath the Spratlys may hold up to 5.4 billion barrels of oil and 55.1 trillion cubic

FIGURE 1.4 Resources of the South China Sea.

feet of natural gas. On top of which, about half of the world's merchant fleet tonnage and nearly one-third of its crude oil pass through these waters each year. They also contain some of the richest fisheries in the world.

REFERENCES

Dahlby, T. (1998). "Crossroads of Asia, South China Sea," *National Geographic*, 194(6): 8–33. National Geographic Society, Washington, DC.

Image Zones maritimes UNCLS.jpg.

Qu et al. (2006). Reprinted with permission of AGU.

Smith, W.H.F. and Sandwell, D.T. (1997). "Global sea floor topography from satellite altimetry and ship depth soundings," *Science*, 277(5334): 1956–1962.

Part II

Tectonics/Geology

2 Geology

2.1 INTRODUCTION

The South China Sea (SCS) and Taiwan Strait were formed primary by the convergence of the Eurasian tectonic plate with two minor lithospheric tectonic plates and a complex boundary belt (Philippine Boundary Belt). The Philippine boundary belt separates the minor lithospheric plates from the Pacific plate. A schematic figure showing both the SCS and the Taiwan Strait is shown in Figure 2.1. These minor plates are the Sunda Plate, Yangtze Plate, and, to a lesser amount, by the Philippine Mobile Belt. A summary of the characteristics of these minor plates and belt is presented in Table 2.1. The convergence of these plates occurred

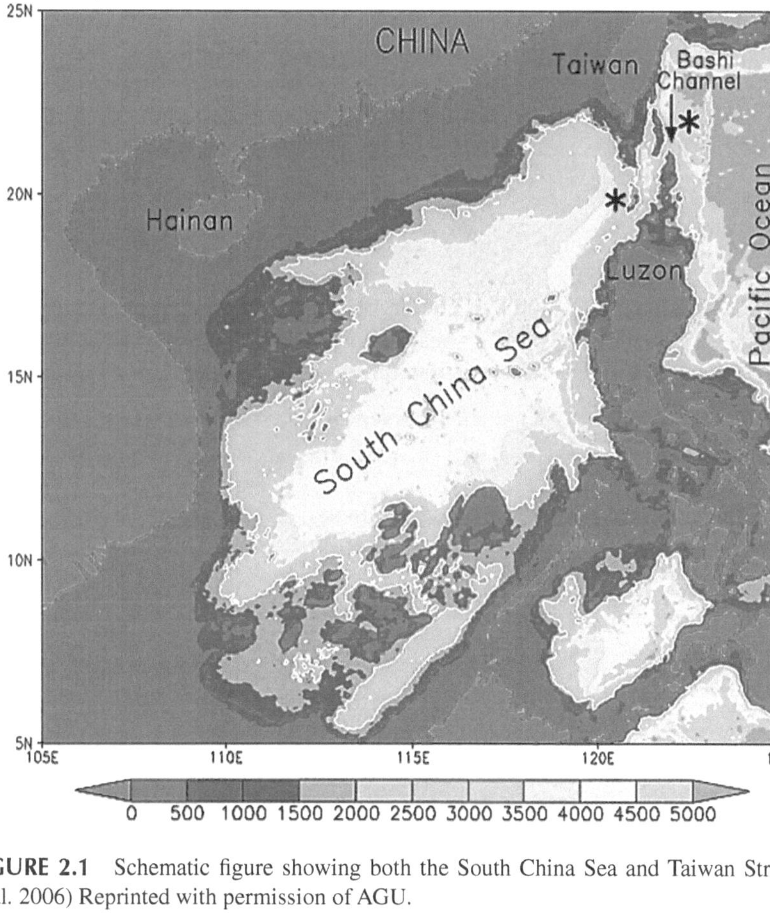

FIGURE 2.1 Schematic figure showing both the South China Sea and Taiwan Strait (Qu et al. 2006) Reprinted with permission of AGU.

TABLE 2.1
Minor Plate Characteristics

Tectonic Plate	Type	Movement/Speed	Features
Sunda	Minor	East/11–14 mm/year	Southeast Asia, Borneo, Java, Sumatra, South China Sea
Yangtze	Minor	South-East/15 mm/year	China
Philippine Mobile Belt	Minor	North-West	Northern Luzon, Philippine Sea
Philippine Plate			

beginning in the Triassic and extended in the Cenozoic. In the following each of these minor plates will be discussed.

2.1.1 Sunda Plate

The Sunda Plate is a minor lithospheric tectonic plate located in the eastern hemisphere along the equator (Figure 2.2). This plate is bounded in the east by the Philippine Plate and the Philippine mobile belt, Molucca Sea Collision zone, Molucca Sea Plate, Banda Sea Plate and Timor Plate; to the south by the Australian Plate to the west; and to the north by the Burma Plate, Eurasian Plate and the Yangtze Plate.

2.1.2 Yangtze Plate

The Yangtze Plate (i.e., South China Block or South China Subplate) comprises the majority of southern China (Figure 2.3). It is bordered in the south by the Sunda Plate and the Philippine Mobile Belt, and in the north and west by the Eurasian and Amur plates. To the east the plate is separated from the Okinawa Plate by a rift that forms the Okinawn Trough. This trough in turn forms a back-arc basin.

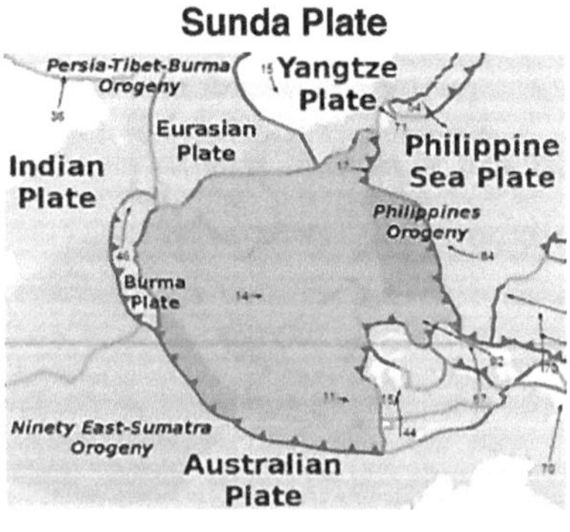

FIGURE 2.2 Sunda Plate (https://en.wikipedia.org/w/index.php?title=Sunda_Plate&oldid=801062099). This file is licensed under the Creative Commons Attribution-Share Alike 4.0 International license.

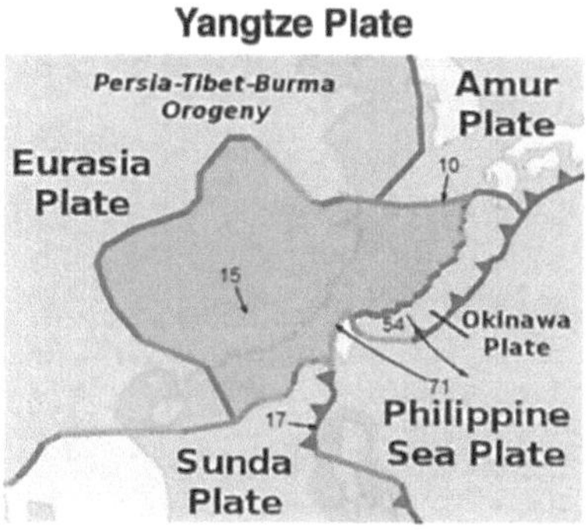

FIGURE 2.3 Yangtze Plate (https://en.wikipedia.org/w/index.php?title=Yangtze_Plate&oldid=824698811). This file is licensed under the Creative Commons Attribution-Share Alike 4.0 International license.

2.1.3 Philippine Mobile Belt

The Philippine Mobile Belt (PMB) is a complex portion of the tectonic boundary between the Eurasian Plate and the Caroline Plate (Figure 2.4). The PMB encompasses most of the Philippines. This belt includes two subduction zones (the Manila Trench to the west and the Philippine Trench to the east) as well as the Philippine

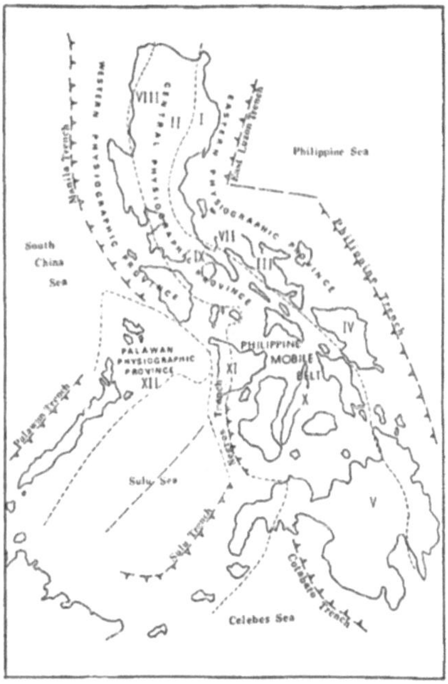

FIGURE 2.4 Major physiographic elements of the Philippine Mobile Belt (https://en.wikipedia.org/w/index.php?title=Philippine_Mobile_Belt&oldid=837442963). This file is licensed under the Creative Commons Attribution-Share Alike 4.0 International license.

Fault System. The belt includes a number of microplates, which have been cut off from the adjoining major and minor plates. These microplates are currently undergoing massive deformation (Galgana et al., 2007).

Northern Luzon, along with a number of other segments comprise the PMB, which is bounded by the Philippine Sea Plate to the east, the Molucca Sea Collision Zone to the south, Sunda Plate to the south-west, and the SCS basin to the west and north-west. To the north the PMB ends in eastern Taiwan, the zone of active collision between the North Luzon Trough portion of the Luzon Volcanic Arc and South China (Schouten and Draut, 2003). Both the eastern coastal range and the inland Longitudinal Valley of Taiwan are formed by accreted portions of the Luzon Arc and the Luzon forearc (Schouten and Draut, 2003).

2.2 SOUTH CHINA SEA GEOLOGY

2.2.1 Topographic Features

The SCS is a marginal ocean basin situated at the junction of the Eurasian, Pacific, and Indo-Australian plates. It is generally agreed that the marginal oceanic basin developed from the Cretaceous period to the Paleogene period. The submarine topography of the Chinese margin is characterized by NE–SW trending broad shelves and narrow slopes.

A schematic of a typical cross-section of the SCS showing the continental crust, oceanic crust, lithosphere, asthenosphere and mantle is presented in Figure 2.5. It was formed by a magma-poor rifting that transitioned to subsequent seafloor spreading in the Oligocene. The area is analogous to the basin and range area in the continental United States. Associated with this marginal ocean basin are (1) sub-basins and (2) island arcs. The basin exhibits the various elements of the Wilson Cycle: continental rifting, seafloor spreading, collision with Borneo to the south, and subduction under the Luzon Arc to the east. The general characteristics of marginal basins is presented in Table 2.2. The seafloor contains Paleozoic and Mesozoic granite and metamorphic rocks. The abysses are caused by the formation of the Himalayas in the Cenozoic. Except for one volcanic island, the islands are made of coral reefs of varying ages and formations. The SCS basin can be divided into three sub-basins based on structural variations (Ding et al., 2016). These sub-basins are the following: East Sub-basin (ESB), the Southwest Sub-basin (SWSB) and the Northwest Sub-basin (NWSB), Figure 2.6. The SWSB is a V-shaped oceanic basin typical of a propagating rift.

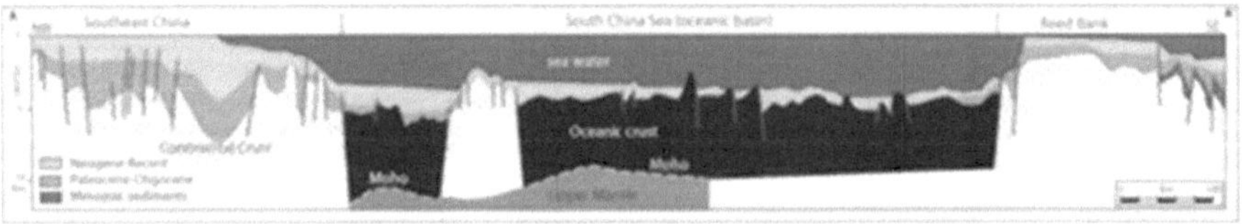

FIGURE 2.5 A north-west to south-east structural transect across the SCS (modified after Yan et al., 2001). Reprinted with permission of Taylor & Francis.

TABLE 2.2
General Characteristics of Marginal Basins

Type of Basin	Morphology	Heatflow	Subduction	Example
Undeveloped	Typical basin	Normal	Commenced and induced convection	Aleutian and Caribbean zones
Active	Active spreading with center generally parallel to axis of island arc	High		Lau-Havre and Mariana Basins
Mature	Widening resulting from spreading	Broad heat flow anomaly		Sea of Japan, Okhotsk and North Fiji basin
Inactive	No spreading occurring	Return to normal	Ceased	South Fiji and West Philippine Basin

Source: Adapted from Toksoz and Bird (1977).

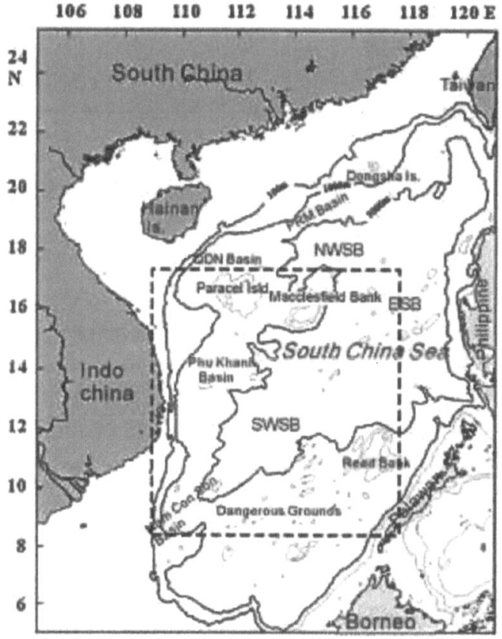

FIGURE 2.6 Major tectonic unit and sedimentary basins in the South China Sea (Ding et al. 2016). Reprinted with permission of Elsevier.

Island arcs occur either when one oceanic plate is subducted beneath another instead of beneath a continent in a manner similar to Andean-type mountain chains. This is similar to the case when two plates meet, and a volcanic island develops on the overriding plate, above the subduction zone marked by a trench. In addition, a small ocean basin usually opens in the far side of the island arc away from the trench. This is called a marginal ocean basin or back-arc basin. The occurrence of sea-floor spreading in an area where two plates are converging is expected to be in a compressional environment. This indicates that the plate motions are associated with a complex interplay of forces and magmatic activity that is not presently understood. A number of island arcs and marginal basins have formed by progressive splitting apart of older arcs in the western Pacific.

Sundaland lies to the west and south-west of the SCS and with a series of volcanic islands to the east. This area is a continental basement/basin consisting of the onshore Indochina Block and the Sunda Shelf that extends to a water depth of 200 m.

The Sunda Shelf consists of stretched continental crust consisting of grabens in the basement rocks overlain by sediment. The continental slope is narrow and steep, and has a water depth that ranges from 200 to 1000 m. In contrast, the oceanic crust extends from water depths of approximately 3000 to over 5000 m. The ocean floor depth in most active marginal basins is similar to spreading ocean-ridge systems. Seismic surveys and gravity inversion data of the SCS indicate that the depth to the Moho (*crustal thickness*) varies from 30 km in the Sunda Shelf to 7 km in the oceanic crust. Dredged rocks in marginal basins have the same geochemistry of either major or minor elements as spreading ridge systems. The marginal basin crust is young based on both direct and indirect evidence: amount of sediment cover, geologic trends, paleomagnetic studies, and fit of predrift continental margins.

The corresponding thickness of the sediment layer on the Sunda Shelf within deep grabens at some locations exceeds 12 km. The Sediment thickness on the oceanic crust is typically 1–2 km. A summary of crustal thicknesses is presented in Table 2.3.

2.2.2 Opening of the South China Sea

The initiation of seafloor spreading in the SCS was established by the International Ocean Drilling Program (IODP) during Leg 349. This organization found that the spreading started at 33 Ma. (Chron 12) early Oligocene, which is the oldest Chron identified in the SCS. This early phase of rifting started in the late Cretaceous to early Paleocene when a Mesozoic convergent margin changed to an extension. Based on magnetic anomaly identification, the end of the SCS spreading could be approximately either 15.5 or 20.5 Ma Miocene (Briais et al., 1993; Barckhausen et al., 2014). Rifting and subsequent seafloor spreading occurred from the Oligocene to the Early Miocene. The seafloor spreading migrated from north-east to south-west.

TABLE 2.3
Crustal Thickness

Location	Crustal Thickness (km)
Sunda Shelf	30
Dangerous Grounds	18–12
Oceanic Crust	11–17

These estimates are based on the Gradstein et al. (2012) geologic time scale (Sorkhabi, 2013). These data further show that the ocean basin opened in at least two major phases. During phase I, the seafloor spreading was north to south, with the spreading ridge striking east–west. This occurred during the period from 30 Ma to ~21 Ma. As a result of this movement a large block was separated from the South China continental block to the north. This block contained the present North Palawan Island, Reed Bank, Macclesfield Bank, and Paracel Islands. At magnetic anomalies 7–6B (between 24.9 and 21.75 Ma) the existing spreading ridge jumped to a north-east–south-west trend. This change in direction resulted in the seafloor spreading in a north-west–south-east direction. This movement eventually separated Reed Bank from Macclesfield Bank.

The chronology of tectonism resulting in rifting is shown in a series of borings from Taiwan to Palawan shown in Figure 2.7 (Franke, 2012). Plotting a graph of age versus latitude for rift-onset unconformity presents an approximate linear relationship (Figure 2.8). A review of Figure 2.8 agrees with the previous observation that the rifting started in the north-east and propagated to the south-west. A large portion of the area near the axis of the east sub-basin is masked by post-spreading magmatic activity (~13–3.5 Ma) (Ishihara and Kisimoto, 1996). This magmatic activity obscures areas of the extinct spreading axis. A map of the magnetic data shows that

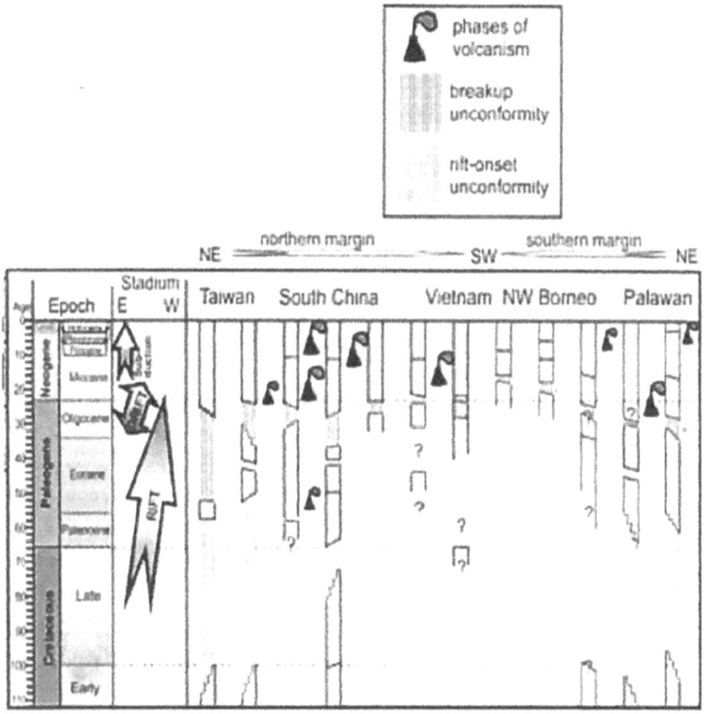

FIGURE 2.7 Chronology of tectonism around the South China Sea. Reprinted with permission of Elsevier.

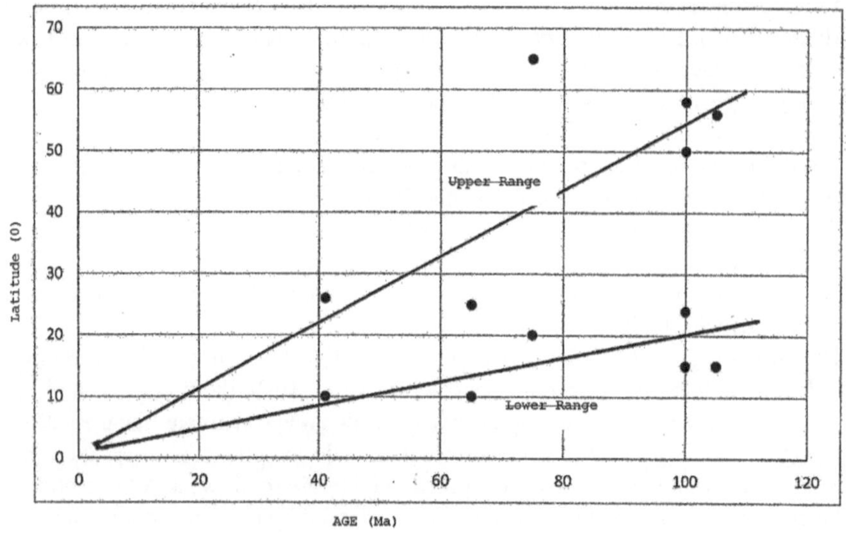

FIGURE 2.8 Breakup unconformity.

a few magnetic lineations belonging to the magnetic seafloor spreading fabric are still preserved. These lineations are parallel to the N055 bathymetric seafloor spreading trends identified on swath-bathymetric maps in the central part of the SCS. This suggests that the extinct ridge axis is N055 trending with N145 transform faults (Figure 2.9) (Sibuet et al., 2016). As a result of this spreading, a number of small islands and banks were separated from the South China continental block. These islands and banks are the following: North Palawan Island, Reed Bank, Macclesfield Bank, and Paracel Islands (Tapponnier et al., 1982). A summary of observed seafloor spreading trends in the SCS has been presented in Figure 2.10.

2.2.2.1 Indian Subcontinent and South-east Extrusion

The Eurasian continent is a basement sedimentary complex that has experienced multiple phases of tectonic deformation. This collision occurred approximately 55–50 Ma based on an estimated velocity of 5 cm/yr for the collision of the Indian plate with the Eurasian continent (Molnar and Tapponier,1976). This resulted in an estimated 2500–2000 km of deformation. The collision of the Indian plate with the Eurasian continent has been simulated in an experiment involving the movement of a rigid block into plasticine material (Sorkhabi, 2013). This experiment indicated that as the rigid block penetrated the plasticine material there was a subsequent lateral movement. As a result of this deformation based on a combination of experimental observations and a review of a tectonic map of Asia there was observed a eastern "flow" of rocks. A large portion of this deformation was taken up by movement on the Red River fault.

Marginal basins are under an extensional stress but very little is known about the opening mechanism. A number of models have been proposed (Sleep and Toksoz, 1971; Bibee et al., 1980). In the Sleep and Toksoz model the subducting lithosphere

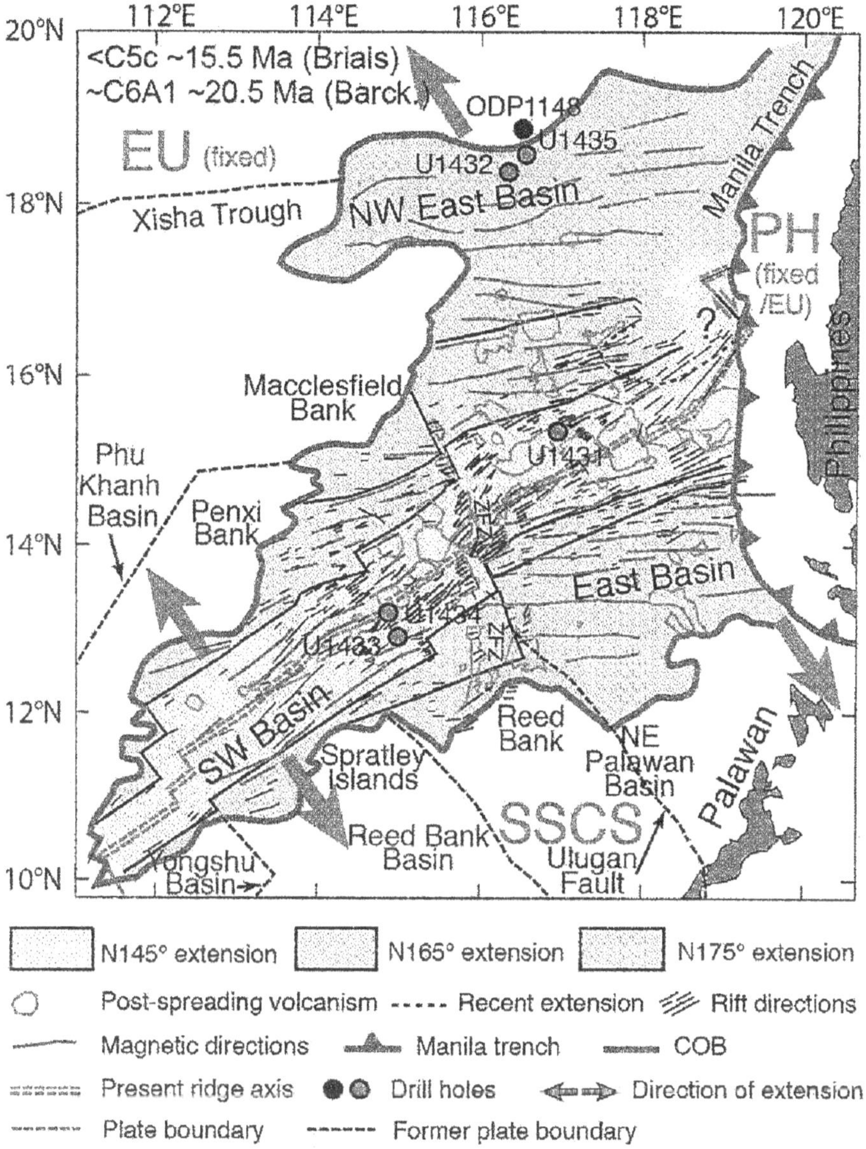

FIGURE 2.9 Seafloor spreading trends in the South China Sea. Black lines bound domains with different spreading trends: the central domain with (U1433, U1434, U1431) N055 trends, the next domain immediately outside the central domain with N075 trends, and the final domain farthest out (East Basin, and NW East Basin) with N085 trends. Large arrows show the direction of extension during the formation of the central domain. The outline of the proposed extinct spreading axis is a contribution of N055 seafloor spreading trends and N145 FZs but is still preliminary. The black dashed lines are features, which acted as plate boundaries during the formation of the SCS. ZFZ, Zhongnan faults zone; EU, PH, and SSCS are Eurasian, the Philippine Sea and southern SCS plates respectively (ref: Sibuet et al. 2016). Reprinted with permission of Elsevier.

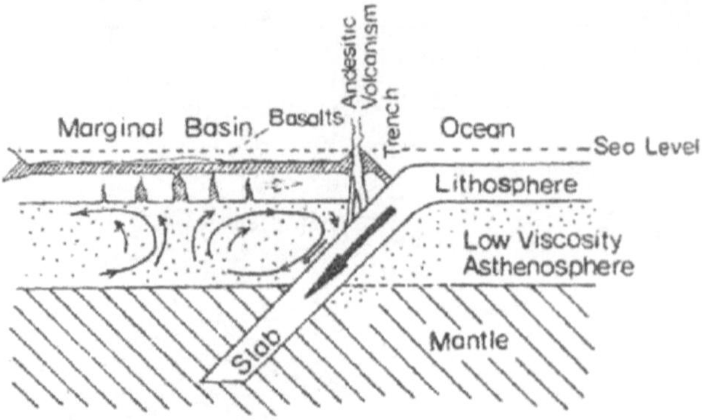

FIGURE 2.10 Model to explain the opening of back arc basins. This is called the secondary flow model (Sleep and Toksoz, 1971). Reprinted with permission of Springer Nature.

drags a part of the low-viscosity asthenosphere into the mantle until the flow is deflected by increasing viscosity and density (Figure 2.10). This movement generates a flow pattern as shown in Figure 2.10 that brings hot asthenospheric material to the base of the lithosphere under the marginal basins. The upwelling of material and subsequent heating of the lithosphere results in a rotating flow pattern that initiates rifting and causing spreading. This theory has been extended by Bibee et al. (1980).

2.2.2.2 Spratly Islands

The Spratly Islands consists of 14 islands ranging in area from 0.4 to 46 ha with an additional 600 coral reefs. A list of the various individual islands is presented in Table 2.4. The largest island is Taiping Island, which is occupied by the Republic of China (ROC). These islands consist of a combination of reefs, banks, and shoals that are built of biogenic carbonate on the higher crests of major submarine horsts. A horst is an uplifted fault block. These horsts are part of a series of parallel and en echelon half-grabens and rotated fault blocks. Their long axis defines linear trends that are parallel to magnetic anomalies exhibited by the oceanic crust. The rotational block faulting is a result of tectonic plates stretching apart. The horsts, grabens, rotated fault blocks, and the rock forming the base consist of stretched and subsided continental crust. This rock is composed of Triassic (220 Ma), Jurassic (192 Ma), and Cretaceous (135 Ma) strata. These rocks include calc-alkalic extrusive igneous rocks, intermediate to acid intrusive igneous rocks, sandstorms, siltstones, dark green claystones, and metamorphic rock. That includes both biotite-muscovite-feldspar-quartz highly complex rock and garnet-mica schists (Wikipedia, 2020).

The total area of the archipelago's islands is 1.77 km^2 (440 acres) and 2 km^2 (490 acres) with reclaimed land.

2.2.2.3 Dangerous Grounds

The Dangerous Grounds is a shallow area in the SCS. The area consists of many low islands, atolls, sand bars, sunken reefs, and reefs rising abruptly from 1000 m water depth.

TABLE 2.4
Various Individual Islands Making up the Spratly Islands

Island Name	In Atoll	Area (ha)	Occupied
Itu Aba	Tizard Bank	46.0	ROC (Taiping Island)
Thitu	Thitu Reef	17.2	Philippines
West York	West York Island	18.6	Philippines
Spratly	Spratly Island	13.0	Vietnam
Northeast Cay	North Danger Reef	12.7	Philippines
Southwest Cay	North Danger Reef	13.0	Vietnam
Sin Cowe	Union Banks	8.0	Vietnam
Nanshan	Nanshan Group	7.9	Philippines
Sand Cay	Tizard Bank	7.0	Vietnam
Loaita	Loaita Bank	6.4	Philippines
Namyit	Tizard Bank	5.3	Vietnam
Amboyna City	Amboyna Cay	1.6	Vietnam
Flat	Nanshan Group	0.6	Philippines
Lankiam Cay	Loaita Bank	0.4	Philippines

Source: Adapted Wikipedia.org/wiki/Spratly Islands.

The Dangerous Grounds area is believed to be a mid-ocean ridge where seafloor spreading started to occur. The total area is approximately 160,000 square miles.

2.2.2.4 Seismicity

The location of major faults in the SCS is presented in Figure 2.9. A review of the seismicity of the area shows that major earthquakes occur only in the immediate area of the Philippines and not in the SCS. In contrast, a number of felt and damaging historical reported earthquakes have occurred in Fujian Province on the mainland in coastal areas.

2.2.2.5 Sediment Transport

A review of Figure 2.11 shows six major rivers discharge into the China Seas (i.e., South China Sea, East China Sea, and Yellow Sea) and their catchment areas. These rivers are the Yangtze, Yellow, Pearl, Red, N. Borneo, and Mekong. The sediment load from these rivers consist predominantly of illite-chlorite clays. In contrast, the Sunda Shelf and Luzon contribute more smectite clays. Milliman and Farnsworth (2011) report that the total sediment load from these rivers is approximately 1900 Mt annually. This river borne sediment is responsible for the development of large deltas and extensive continental shelves in the China Seas.

The source of the large sediment supply from the above rivers is a function of a number of factors. These factors are the topography, climate, geology of the drainage basins, and human activity. The sources of the Yangtze, Yellow, and Mekong rivers are all located in the Tibetan Plateau. Table 2.5 lists the three major rivers (Pearl, Mekong, and Red) discharging into the SCS and their characteristics. A review shows that these three rivers discharge an estimated 360 Mt of sediment per year.

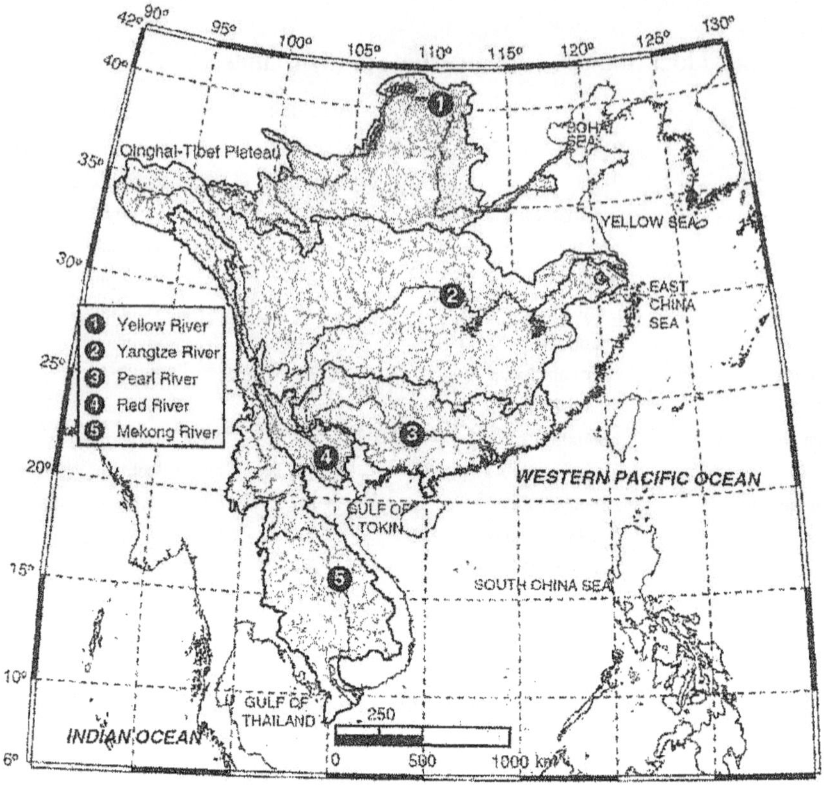

FIGURE 2.11 Five major rivers discharging into the South China Sea and their catchment areas (Wang et al., 2011). Reprinted with permission of Elsevier.

TABLE 2.5
The Three Major Rivers Discharging into the South China Sea

River	Area (10^3 km^2)	Length (km)	Max. Elev. (km)	Ann. Prec. (mm)	Water Disch. (km^3/yr)	Sed. Disch. (Mt/yr)	Delta Area (10^3 km^2)
Pearl	440	2100	1800	1600–2300	300	70	41
Red	150	1100	2200	1120	120	130	14
Mekong	810	4400	5400	1300	500	160	93
Totals				4020–4720	920	360	148

Source: Adapted from Wang et al. (2014).

2.3 TAIWAN STRAIT GEOLOGY

2.3.1 Topographic Features

The Taiwan Strait is located on the south-east coast of East Asia, connecting to the East China Sea in the north-east and to the SCS in the south-west. The strait extends from an imaginary north-east line from Pingtan Island in Fujian Province to Fukueicuiao Cape north of Taiwan island. In the south-west the boundary is an imaginary line extending from Nan-Ao Island in Guangdong province to Oluanpi Cape south of Taiwan Island. The strait therefore extends over a length of 375 km in a NE–SW direction. The width of the strait at its narrowest distance is approximately 130 km from Pingtan Island to Hsinchu on Taiwan Island at the northern end and rather broad at its southern end. The Taiwan Strait is a part of the continental shelf of the East China Sea located on the south-east edge of the Eurasian plate, behind the collision zone of the Philippine Sea plate with the Eurasian plate. It is a frontal basin of the plate. Taiwan Strait earthquake shocks are characteristic of a typical intraplate earthquake.

A bathymetric map of the strait is presented in Figure 2.12. A review of this figure shows the presence of both shallow and wide shelves covered by water in the East China Sea, the Taiwan Strait, and the SCS along the southeastern Chinese margin. The Huatung Basin with water depths between 4000 and 5000 m lies off the eastern Taiwan coast. North-east of Taiwan lies the back arc basin of Okinawa. The deepest submarine feature of this basin is the Ryukyu Trench, which is south of the Ryukyu Arc with a minimum water depth of more than 6000 m. The deep oceanic Huatung basin is bounded by the Gagua Ridge and N–S trending ridges and troughs with irregular surface characterize the seafloor off southwestern Taiwan. Off south-eastern Taiwan a broad submarine slope dips down to 3000 m in water depth. This slope is dissected by numerous canyons juxtaposed with the SCS slope.

Taiwan is situated between the Ryukyu and Philippine Islands (Figure 2.13a). Unlike other islands in the vicinity it is in a collision regime (Biq, 1972; Wu, 1978; Lin and Tsai, 1981; Ho, 1982; Tsai, 1986; Wu and Chingchang, 1991). This collision has resulted in intense seismicity, relatively high mountain ranges composed of Paleogene and Neogene sediments along the axis of the island and rapid uplifting of marine terraces in the recent past (Wu and Chingchang, 1991). The continued collision induced in Taiwan involves the passive margin of the Eurasian plate off the south-east Chinese coast and the northern end of the Luzon–Taiwan arc that is associated with the Philippine Sea plate (Figure 2.13a).

The east–west profile of Taiwan Island as shown in Figure 2.13b consists of a coastal range composed of Miocene andesites and younger sedimentary rocks. Andesite is a gray volcanic rock of intermediate composition between basalt and rhyolite. The coastal range is a former island arc that can be linked to active andesitic islands off the south-eastern coast of Taiwan. Next along the east–west profile is the "long valley (longitudinal valley)," which is a prominent geomorphic feature. Further to the west in the central range are Pliocene clastics and Quaternary molasses. Holocene sediments cover the western Coastal Plain and foothills. In the north-south profile young andesitic volcanoes are found in the extreme north of the island.

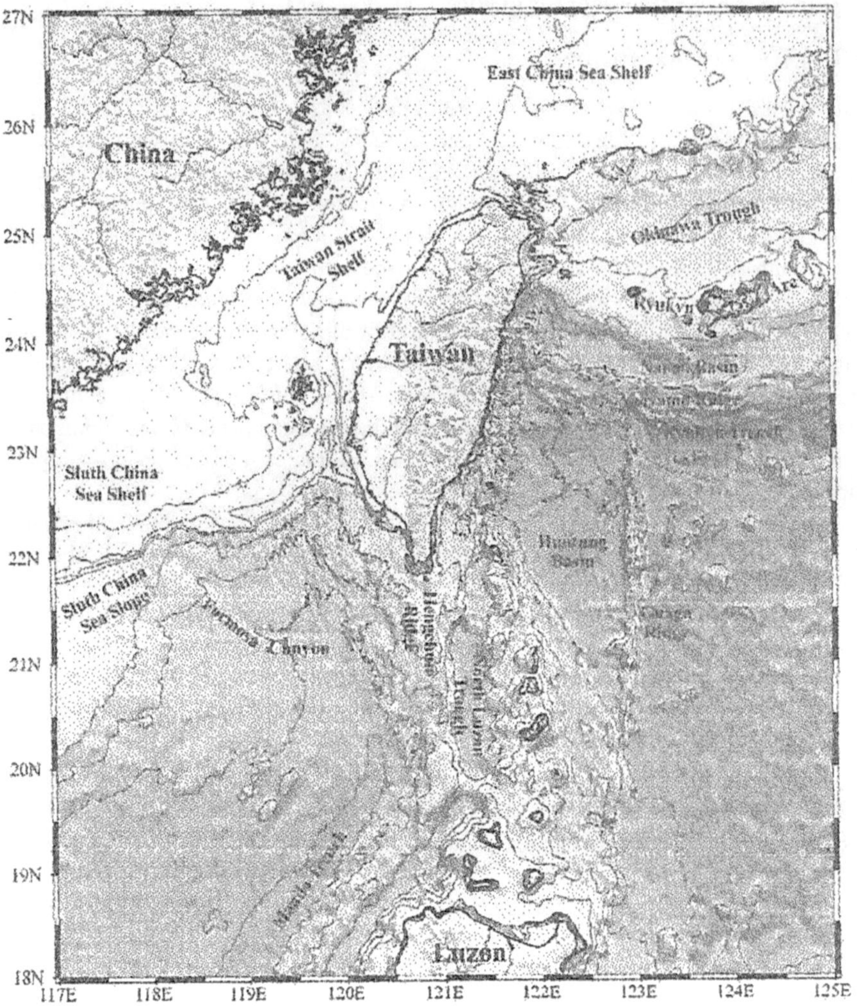

FIGURE 2.12 Bathymetric map of Taiwan Island and Strait (Yu, 2003). Reprinted with permission of Taylor & Francis.

In addition, in the Miocene-Pliocene coarse-grained sandstone clastics is often found along with coal above North 23 latitude while rocks of the same age in the south are fine-grained and contain little coal, Wu and Chingchang (1991).

2.3.2 Taiwan Strait Tectonics

Taiwan Strait is a depression basin on the foreland of the Eurasian plate as shown in Figure 2.14. It is also a kind of continental shelf bounded and separated by a major NW transform-like fault, the South China Shelf and by several NW faults in the north from the East China Sea Shelf. According to Peng et al. (2003), the large NW faults

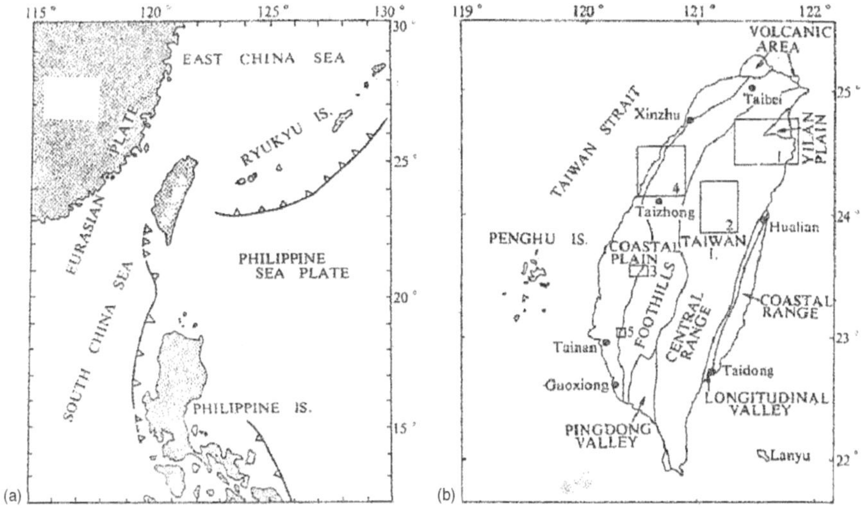

FIGURE 2.13 (a) Plate tectonic environment and the location of Taiwan, (b) Physiographic provinces and major towns in Taiwan (Wu and Chingchang, 1991).

mostly belong to a kind of differentiating boundary between geologic bodies, similar to Peikang Basement High.

The topographic and geomorphologic features of the Taiwan Strait make it a fundamentally gentle, shallow shelf with an average depth less than 60 m. To its west is the down-flexing foreland of the Eurasian plate bordering the NE great deep offshore fault of Fujian, and to its east is the foreland deep bordering on Taiwan orogen by overthrust.

Marine geology in the Taiwan region reflects a transition from a passive margin to an active margin during the last 5 million years. Taiwan Island was formed by the collision of the Luzon Volcanic Arc with the Chinese continental margin.

2.3.3 Seismicity

The location of major faults in Taiwan and the Taiwan Strait is presented in Figure 2.14. The seismicity of the area is shown in Figure 2.15. A review shows that major earthquakes occur on Taiwan and not in the strait. This conclusion is also verified by no major earthquakes being reported on instruments in the Taiwan Strait. In contrast, a number of felt and damaging historical reported earthquakes have occurred in Fujian Province on the mainland in coastal areas (Figure 2.16). The largest reported earthquake took place in 1604 near Quanzhow causing severe damage. Based on intensity distribution diagrams, the epicenter was probably located in the strait (Committee on Fujian Earthquakes, 1979). No effects of this earthquake were recorded in Taiwan because records were not kept until the mid-1600s. The locations where historically large earthquakes are known to occur (i.e., Fuzhou, Shantou and Quanzhou) all show evidence of recent faulting.

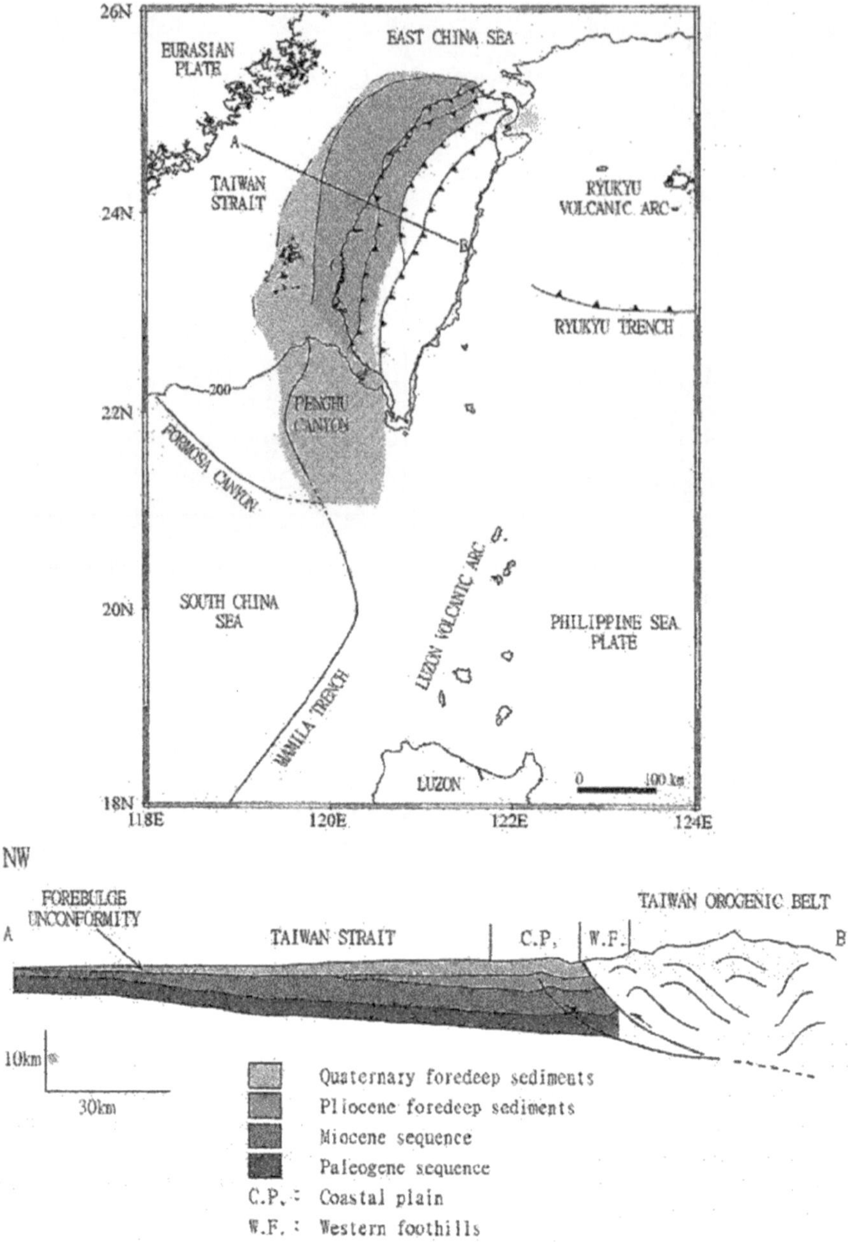

FIGURE 2.14 Geological setting in the Taiwan region. Reprinted with permission of Taylor & Francis.

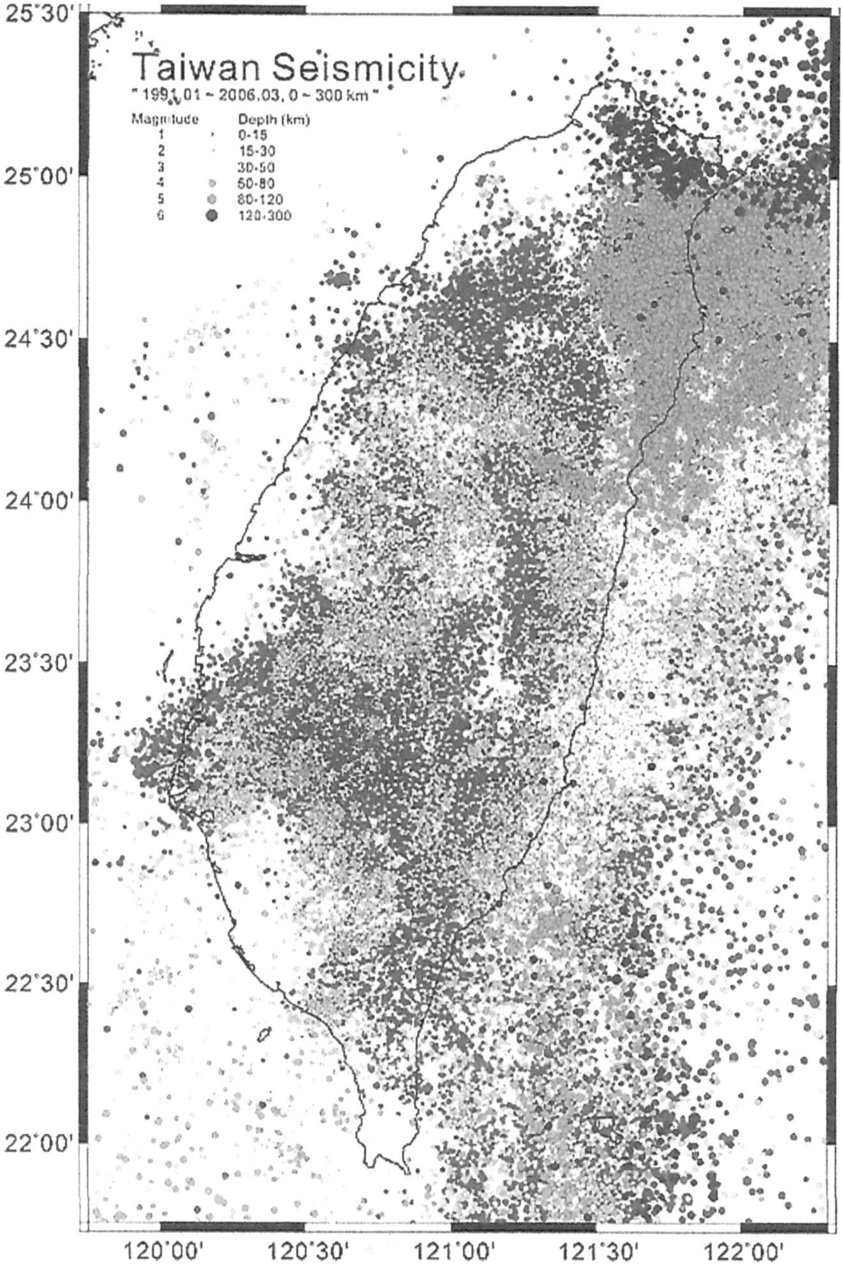

FIGURE 2.15 Taiwan seismicity showing both magnitude and depth of earthquakes (Hsu Shihhung, Chinese Wikimedia Commons). This file is licensed under the Creative Commons Attribution-Share Alike 2.5 Generic (https://creativecommons.org/licenses/by-sa/2.5/deed.en) license.

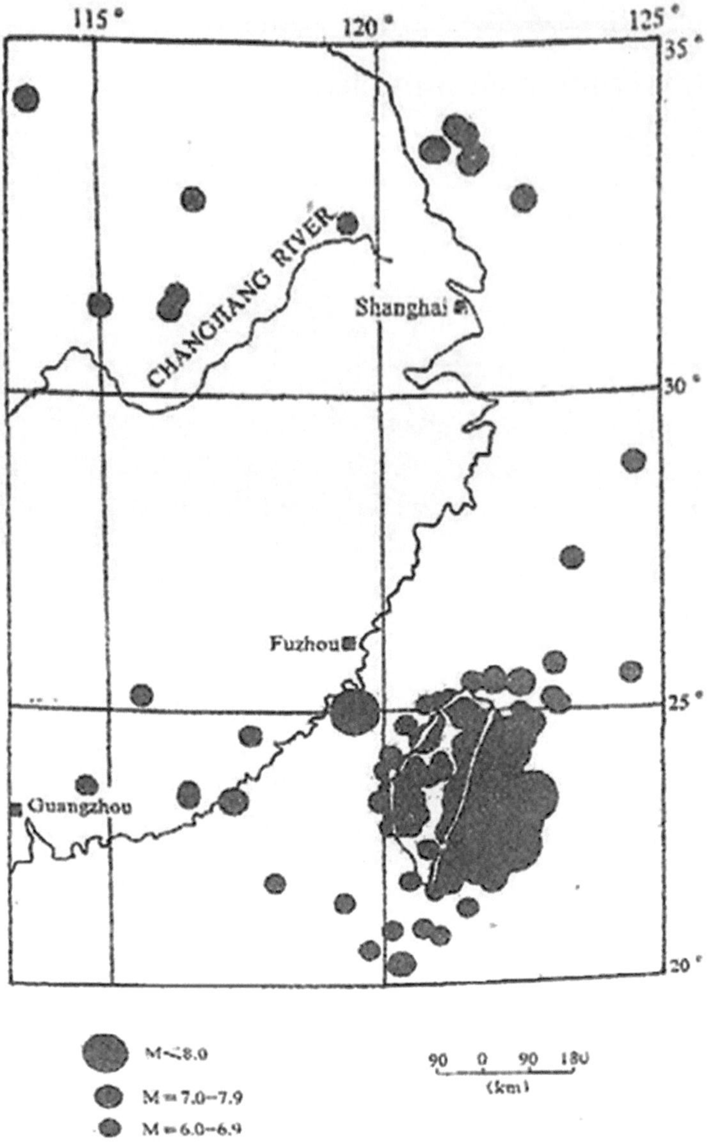

FIGURE 2.16 Major earthquakes in Taiwan and south-eastern China (Modified from Institute of Geophysics, 1976).

2.3.4 Sediment Transport

The distribution of Taiwan mountainous rivers and their annual sediment loads to the surrounding seas is presented in Figure 2.17. This figure shows that approximately 70 Mt/yr of sediment are transported per year into the SCS, 60–150 Mt/yr into the Taiwan Strait, and 150 Mt/yr into the Pacific.

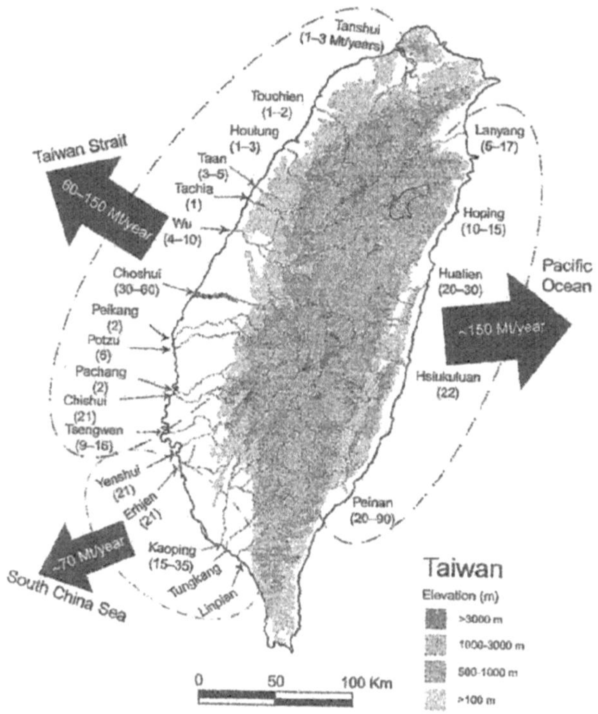

FIGURE 2.17 Distribution of Taiwan mountainous rivers and their annual sediment loads to the corresponding seas (Liu et al., 2007). Reprinted with permission of Elsevier.

REFERENCES

Barckhausen, U., Engels, M., Franke, D., Ladage, S., and Pubellier, M. (2014). "Evolution of the South China Sea: Revised ages for breakup and seafloor," *Marine Petroleum Geology*, 58: 599–611.

Bibee, L.D., Shor, G.G., Jr., and Lu, R.S. (1980). "Inter-arc spreading on the Mariana trough," *Marine Geology*, 35: 183–196.

Biq, C.C. (1972). "Transcurrent buckling, transform faulting and transpression: Their relevance in eastern Taiwan kinematics," *Petroleum of Geology Taiwan*, 10: 1–10.

Briais, A., Patriat, P., and Tapponnier, P. (1993). "Updated interpretation of magnetic anomalies and seafloor spreading stages in the South China Sea: Implication for the tertiary tectonics of Southeast Asia," *Journal of Geophysical Research*, 98(B4): 6299–6328.

Ding, W., Li, J., and Clift, P.D. (2016). "Spreading dynamics and sedimentary process of the Southwest Sub-basin, South China Sea: Constraints from multi-channel seismic data and IODP Expedition 349," *Journal of Asian Earth Sciences*, 115(January): 97–113.

Franke, D. (2012). "Rifting lithosphere breakup and volcanism: Comparison of magma-poor and volcanic rifted margins," *Marine and Petroleum Geology*, 43: 63–87.

Galgana, A., Hamburger, W., McCaffrey, R., Bacolcol, C., and Aurella, M. (2007). "Modelling the Philippine Mobile Belt: Tectonic blocks in a deforming plate boundary zone," AGU (American Geophysical Union) Fall Meeting December 2007, http://adsabs.harvard.edu/abs/2007AGUFM. G21C0670G.

Girton, J.B., and Whitehead, I.A. (2006). "Deepwater overflow through Luzon Strait," *Journal of Geophysical Research*, 111.

Gradstein, F., Ogg, J., and Smith, A.G. (2012). *A Geologic Time Scale 2004*, Cambridge University Press, Cambridge, https://doi.org/10.1017/CB09780511536045.

Ho, C.S. (1982). *Tectonic Evolution of Taiwan: Explanatory Text of the Tectonic Map of Taiwan*. Ministry of Economic Affairs, Taiwan, 126pp.

Ishihara, T., and Kisimoto, K. (1996). "Magnetic anomaly map of east Asia" (14,000,000, CD ROM Version), Tokyo, Japan Geological Survey of Japan, AIST.

Lin, M.T., and Tsai, Y.B. (1981). "Seismotectonics in the Taiwan-Luzon area," *Bull. Inst. Earth Sci., Acad. Sci.*, 1: 54–82.

Liu, Y., Santos, A., Wang, S.M., Shi, Y, Liu, H., Yuen, D.A. (2007). "Tsunami hazards along Chinese coast from potential earthquakes in South China Sea," *Physics Earth and Planetary Interiors*, 163: 233–244.

Milliman, J.D., and Farnsworth, K.L. (2011). *River Discharge to the Coastal Ocean: A Global Synthesis*. Cambridge University Press, New York, 384pp.

Molnar, P., and Tapponnier, P. (1976). "The collision between India and Eurasia," *Scientific American*, 236(4): 30–41.

Peng, F., Ye, Y., and Pan, G. (2003). "Major features of topography, geology, and crustal stability in the Taiwan Strait: A scientific approach to the problem of the Taiwan Strait Tunnel Project," *Marine Georesources and Geotechnology*, 21: 121–138.

Schouten, C., and Draut A.E. (2003) "A general model of arc-continent collision and subduction polarity reversal from Taiwan and the Irish Caledonides," in R.D. Larter, and P.T. Leat (Eds.), *Intra-oceanic Subduction Systems: Tectonic and Magmatic Processes*, Geological Society, London, pp. 81–96, Special Publications, 219pp.

Sibuet, J.C., Yeh, Y.-C., and Lee, C.-S. (2016). "Geodynamics of the South China Sea," *Techonophysics*, 692(Part B): 98–119.

Sleep, N., and Toksoz, M.N. (1971). "Evolution of the Marginal Basins," *Nature*, 233: 548–550, Macmillan Journals Ltd.

Sorkhabi, R. (2013). "South China Sea Enigma," *GEOPro*, 10(1): 13.

Tapponnier, P., Peltzer, G., Le Dain, A.Y., Armijo, and Cobbold, P. (1982). "Propagating extrusion tectonics in Asia: New insights from simple experiments with plasticine," *Geology*, 10: 611–616.

Toksoz, M.N., and Bird, P. (1977). "Formation and evolution of marginal basins and continental plateaus," in M. Talwani, & W. C. Pitman III (Eds.), *Island Arcs, Deep Sea Trenches and Backarc Basins*. American Geophysical Union, Washington, DC, pp. 379–393.

Tsai, Y.B. (1986). "Seismotectonics of Taiwan," *Tectonophysics*, 125: 17–37.

Wang, P., Li, Q., and Li, C.F. (2014). *Geology of the China Seas*, Elsevier, Burlington, 687pp.

Wang, H., Saito, Y., Zhang, Y., Bi, N., Sun, X., and Yang, Z. (2011). "Recent changes in sediment flux to the western Pacific Ocean from major rivers in east and southeast Asia," *Earth Science Reviews*, 108: 80–100.

Wu, F.T. (1978). "Recent tectonics of Taiwan, *Journal of the Physics of the Earth*, 26: 265–299.

Wu, F.T., and Chingchang, B. (1991). "Collison-induced extension tectonics in Taiwan and Fujian, China," in *Proceedings of Symposium on Geology and Sedimology of the Taiwan Strait and Its Coasts*, China Ocean Press, United Kingdom, pp. 318–331.

Yan, P. et al. (2001). "A north-west to south-east structural transect across the SCS," *Tectonophysics*, 386.

Yu, H-S. (2003). "Geological characteristics and distribution of submarine physiographic features in the Taiwan region," *Marine Georesources and Geotechnology*, 21(3–4): 139–155.

Wikipedia. (2020). Spratly Islands, https://en.wikipedia.org/wiki/Spratly_Islands

Part III

Oceanographic Factors

3 Oceanographic Factors

3.1 INTRODUCTION

A study of oceanographic factors includes driving mechanisms to create flow of water in the ocean (waves, tsunamis, tides, horizontal, and vertical currents) and sea water chemical properties. These primary energy sources are due either to radiation or gravitational attraction. The effect of these energy sources on a body are either direct or indirect.

Flow of water in the oceans is created by various mechanisms. These mechanisms can be broadly divided into those resulting from a direct application of forces and those from an indirect application. Direct forces are those resulting from astronomical influences. These are (1) solar and lunar attraction, and (2) the uneven heating of the earth's surface. These forces result in tides, tide-induced currents, and currents driven by density gradients. In contrast, the indirect forces can be described as either impulsive forces or meteorological forces. These forces result in tsunamis, turbidity currents, waves, wave-induced currents, storm tides (surges), and currents. A summary of these flow mechanisms has been presented by Wiegel (2013).

This chapter is organized into three major sections. These sections are: (1) Earth System Processes, (2) Tides, and (3) Currents.

3.2 EARTH SYSTEM PROCESSES

The radiant energy of the sun passes through the atmosphere before reaching the world's oceans. The atmosphere is basically transparent to light, and we tend to assume that this condition exists for all electromagnetic radiation. In fact, the gases comprising the atmosphere selectively scatter light depending on their wavelengths. These gases also absorb electromagnetic energy at specific wavelength intervals called absorption bands. The intervening regions of high-energy transmittance are called atmospheric transmission bands, or windows. The transmission and absorption bands through the atmosphere are shown in Figure 3.1, together with the gases responsible for the absorption bands.

A review of Figure 3.1 shows that the ozone (O_3) layer in the upper atmosphere is completely absorbed at wavelengths less than 0.3 μm. In clouds, water occurs as aerosol sized particles of liquid at wavelengths shorter than 0.3 μm rather than as vapor. Electromagnetic radiation at wavelengths less than about 0.3 μm are absorbed and scattered in clouds. Only radiation of microwaves and longer wavelengths is capable of penetrating clouds without being scattered, reflected, or absorbed.

The relationship between wavelengths of electromotive force (*emf*) and the amount of black body energy radiated as a function of temperature expressed in degrees Kelvin (°K − 273 = °C) is shown in Figure 3.2. The sun, with a surface

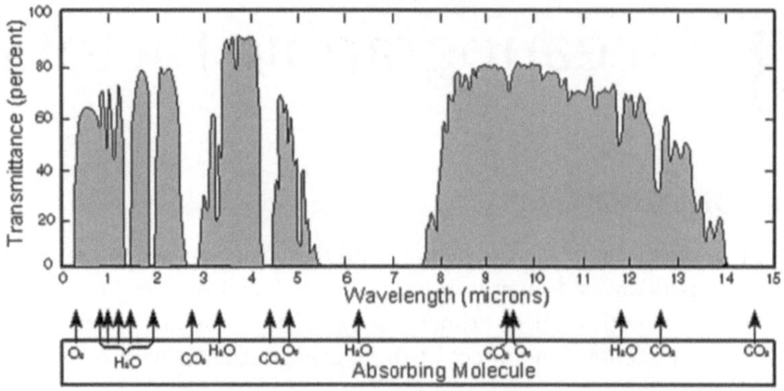

FIGURE 3.1 Transmission of energy through the atmosphere as a function of wavelength. Wavelength regions of high transmittance are atmospheric windows. US Government, Public Domain.

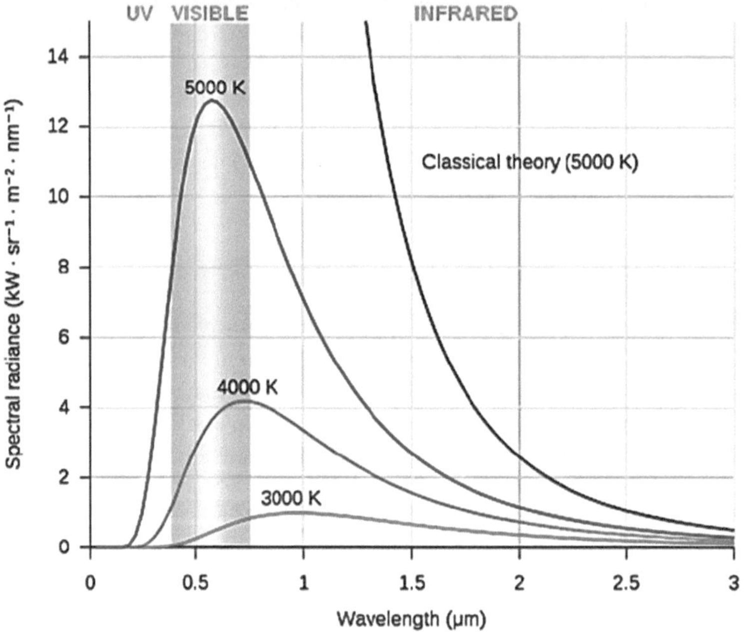

FIGURE 3.2 Spectral distribution curves of energy radiated from objects at different temperatures (https://upload.wikimedia.org/wikipedia/commons/thumb/1/19/Black_body.svg/1200px-Black_body.svg.png).

temperature of almost 6000°K, radiates enormous amounts of energy at wavelengths all across the ultraviolet, visible, and infrared bands with the maximum concentration of energy occurring at a wavelength of about 0.5 μm, corresponding to green light as shown in a spectrum of a spectrum (Figure 3.2). Thus during daylight hours the maximum energy incident on the earth and reflected from the earth is in the visible band with a maximum energy peak at a wavelength of 0.5 µm. The average surface (ambient) temperature of the earth is 290°K (17°C). The energy radiated from the earth at this temperature is distributed as a broad low curve in the infrared band with a peak concentration at 8.7 µm. This radiant energy level is very low in comparison to reflected solar energy, but is dominant at night in the infrared band.

The earth retains 19% of the total energy falling in the atmosphere. This energy is directly absorbed by atmospheric dust, water vapor, ozone, and a small amount by clouds. A large amount of energy is also scattered back into space from clouds with smaller amounts back-scattered from air, dust, haze in the atmosphere, and from the sea surface. The remaining energy, about 52%, penetrates to the earth's surface. Of this amount approximately 70% is water and is absorbed as heat.

The temperatures of the atmosphere and surface ocean water have remained relatively constant over time due to the energy absorbed by the atmosphere and surface layers of the ocean. The long-term steady state exists because the total amount of thermal and radiant energy leaving the planet just balances the solar energy incident on the ocean, land, and atmosphere. The energy released from the earth is primarily in the form of long-wave radiation from clouds, water vapor, $CO_2\mu$, and a small amount that is radiated directly from the ocean to space.

The majority of long-wave (infrared) energy radiated from the ocean is converted to heat in the lower atmosphere. This heat is transmitted from the ocean by a combination of conduction and convection to the lower atmosphere. Heat is also added to the latent heat. The latent heat is involved in the evaporative transfer from the ocean surface to the lower atmosphere. This results in the ocean being heated from above while, the atmosphere is heated from below, Figure 3.3. This leads

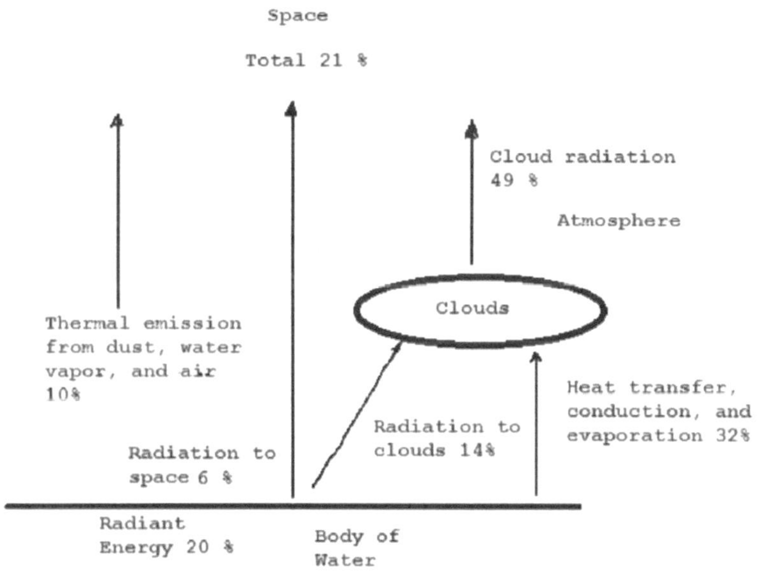

FIGURE 3.3 Energy flux from the ocean and atmosphere. Transfer of heat and emission of long-wave (thermal infrared and microwave) radiation.

to stability in the upper atmosphere and instability in the lower atmosphere (Figure 3.3). The instability leads to atmospheric circulation.

3.3 WINDS AND STORMS

The earth is surrounded by an envelope of gases approximately 62 miles in thickness. The effect of the rotation of the sphere and the sun's radiation on this gaseous envelope will be discussed next. The effects of the sun's radiation on a stationary sphere will be considered first. This will be followed considering the effect on a rotating sphere.

3.3.1 Stationary Earth

Assume that the earth can be modeled as a sphere. This sphere is motionless and covered with layers of water (i.e., oceans) and air (i.e., atmosphere). The lower portions of the air are in contact with the water and are therefore saturated with water vapor. The air and the water are then heated by solar radiation. The heat due to solar radiation becomes more intense near the equator. The heating at the equator is more intense because it directly faces the sun while areas above and below are at a slant. The air in the vicinity of the equator is heated relative to air further from the equator; its density will decrease, and it will therefore rise. This movement will cause surface air from higher latitudes to flow toward the space left by the rising equatorial air. In reality, the rising air represents a region of relatively low atmospheric pressure. Air, like any fluid, flows from an area of high potential to an area of low potential. The heated equatorial air rises to an altitude determined by its density and begins to spread laterally toward the poles. As equatorial air rises into a region of lower atmospheric pressure it expands, its temperature decreases, and it becomes saturated with water vapor to the extent that condensation occurs. This idealized world is shown schematically in Figure 3.4. The equatorial surface regions of the globe are characterized by warm temperature, relatively low atmospheric pressure, clouds, and rainfall. In the polar regions, the temperatures are low, the surface atmospheric pressure is high, and water vapor content is low, which causes precipitation to be low.

3.3.2 Rotating Earth

The theoretical effect of the earth's rotation on the pattern of wind at the surface of the ocean is considered next. The circumference of the earth decreases to toward the poles. Every object attached to the earth as it spins counter clockwise on its axis moves to the east at a velocity dependent on latitude. The deflection of a moving object on a rotating earth is caused by Coriolis affect (refer to Figure 3.5a).

The angular velocity caused by this spinning is highest at the equator. In contrast, the angular velocity decreases as you move away from the equator. This decrease in the rate of angular velocity with increasing latitude is not uniform. An idealized depiction of atmospheric circulation based primarily on the above is presented in Figure 3.5b.

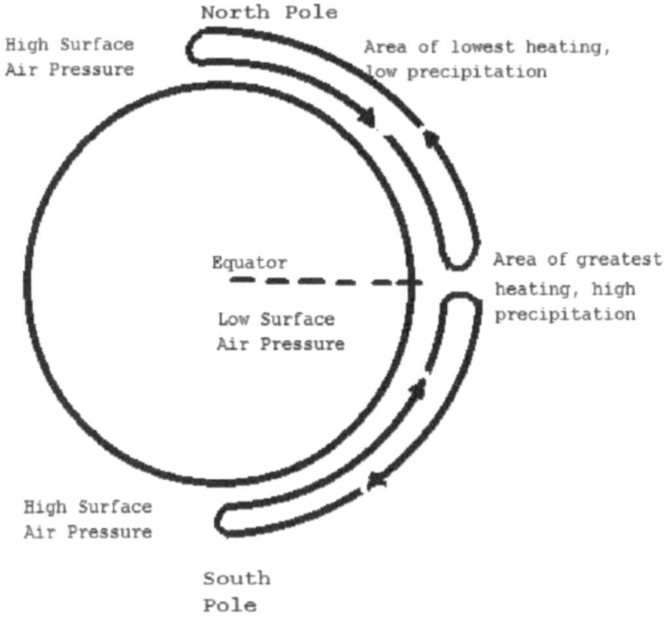

FIGURE 3.4 Atmospheric circulation of a non-rotating earth.

3.3.3 South East Asia Wind Pattern

The climate of the SCS is dominated by the monsoonal winds rising out of the differences in the heat-absorbing capacity of the continent and the ocean. Although both the continent and the ocean receive the same amount of heat from the sun, the former warms up and cools faster than the ocean. In winter, the air over the continent is cooler than that over the ocean. The high air pressure in winter is formed in the Asian hinterlands where heat radiation from the winter sun is weak and a large amount of heat is lost in the long night, rendering the air very cold. When sufficient cold air accumulates, it moves southward in a howling north or north-west wind that causes a sharp drop in temperatures resulting in a cold wave. The greater part of the country is therefore cold and dry in winter, making China a comparatively cold country among those of its latitude. In summer, an area of high air pressure is formed over the ocean and that subsequently moves toward the continent. The greater part of China has then south-east and south-west winds. As they pass across the vast ocean surface, the summer monsoon winds picks up moisture and releases it over the Chinese mainland. This is why the annual precipitation over the greater part of China is highly concentrated within a few summer months. A result of this movement of air masses is the development of surface currents. A schematic illustration of this movement is presented in Figure 3.6.

3.3.4 Global Wind Pattern

Large-scale movement of air distributes thermal energy on the earth's surface. This movement of air is due to solar radiation (i.e., heating) and the laws of thermodynamics.

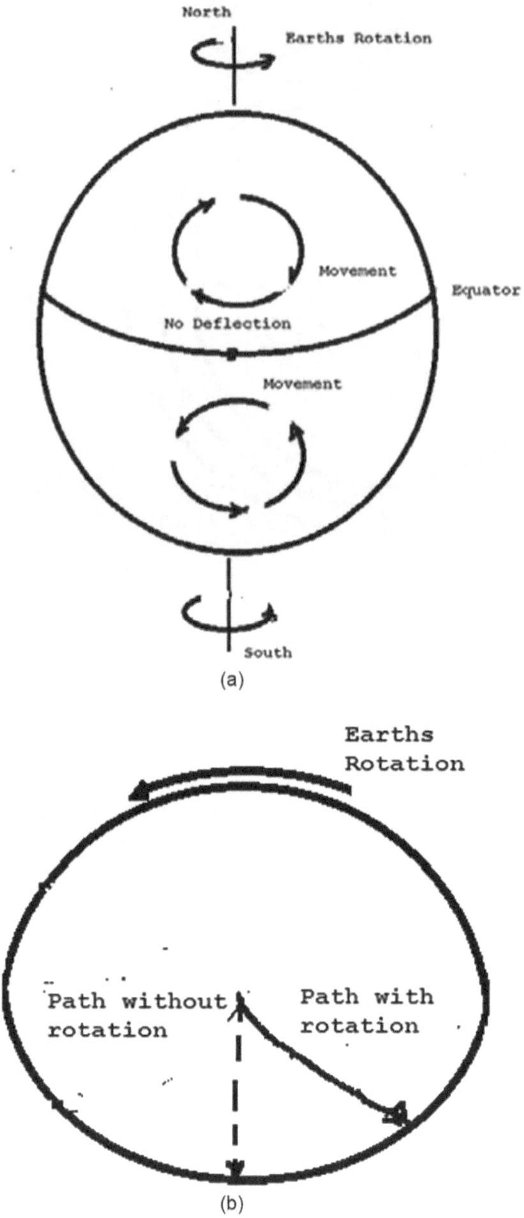

FIGURE 3.5 The deflection of liquids and gases due to the Coriolis effect.

This heating results in a decrease in the density of air causing air masses to move. The resulting atmospheric circulation is relatively constant as shown in Figure 3.7. Deviations from this circulation pattern occur when the circulation cells (Hadley cells) shift polewards in warm periods (i.e., inter-glacial) and towards the equator during cold periods (i.e., glacial).

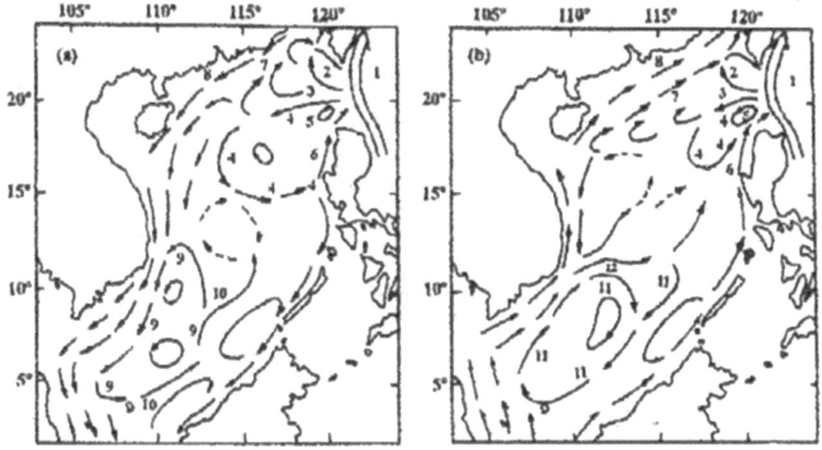

(a) (b)

FIGURE 3.6 Schematic diagram of the SCS circulation patterns (a) Winter and (b) Summer. *Notes*: 1. Kuroshio, 2. Loop current, 3. SCSHK, 4. Luzon Cyclonic Gyre, 5. NW Luzon Cyclonic Eddy, 6. NW Luzon Coastal Current, 7. SCSWC, 8. Guangdong Coastal Current, 9. SCS Southern Cyclonic Gyre, 10. Natuna Offshelf Current, 11. SCS Southern Anticyclonic Gyre, 12. SE Vietnam Offshore Current (After Fang et al., 1998).

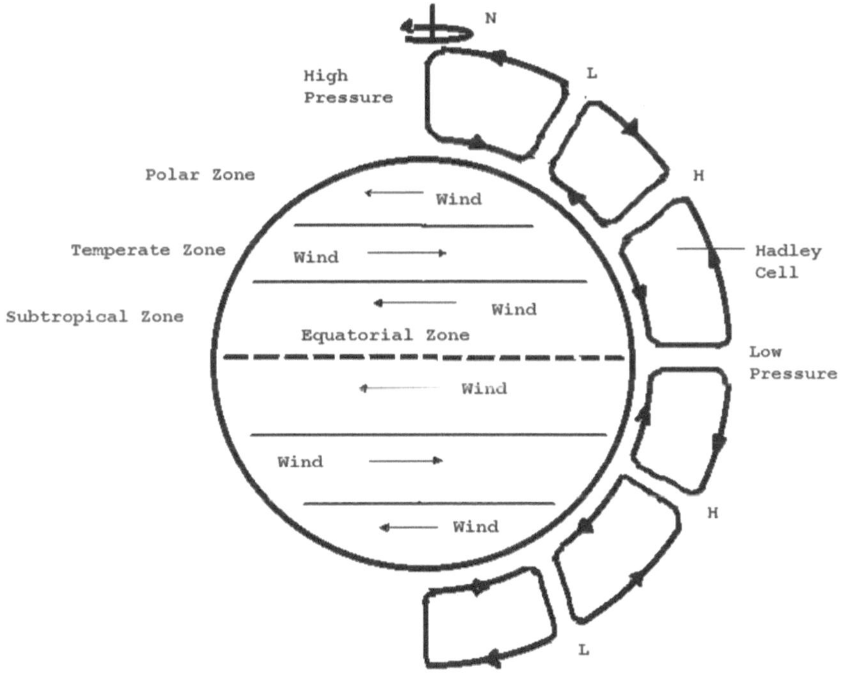

FIGURE 3.7 Idealized depiction of atmospheric circulation (Adapted from Wikimedia Commons).

3.4 TIDES

3.4.1 Origin

Tides are the periodic rise and fall of sea level in response to the gravitational attraction of the sun and moon. This motion is modified by the rotation of the earth, friction forces, and the ocean boundaries. A summary of tidal datum is presented in Figure 3.8. Significant motions of the earth–moon–sun system include the revolution of the earth about the sun and the revolution of the moon about the earth. These motions describe orbits that are approximately elliptical in form. In addition, the moon and earth each rotate about their own axes. The plane in which the earth revolves about the sun is called the ecliptic plane; the axes of the earth is inclined at 66.5° to the ecliptic plane. The moon's orbit about the earth is inclined at 5° 9′ to the ecliptic plane (Figure 3.9). The sun and earth are each displaced a small distance from the centers of their respective orbits. The positions of the moon when it is nearest to and farthest from the earth are called the perigee and apogee, respectively. At perigee the moon has its greatest tide-producing effect on the earth. In contrast, the nearest and farthest positions of the earth from the sun are called the perihelion and aphelion, respectively. A schematic illustration showing these relationships is

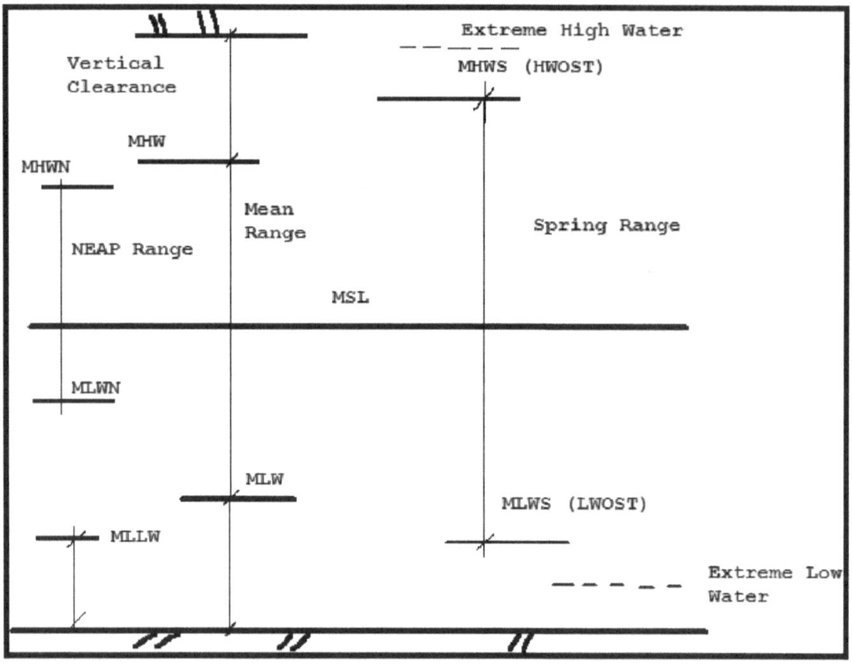

FIGURE 3.8 Summary of Tidal Datum. *Notes*: MLW – mean low water; MLLW – mean lower low water; MSL – mean sea level; MHW – mean high water; MLWN – mean low water neap; MHWN – mean low water neap; MHWS – mean high water spring; MLWS – mean low water spring.

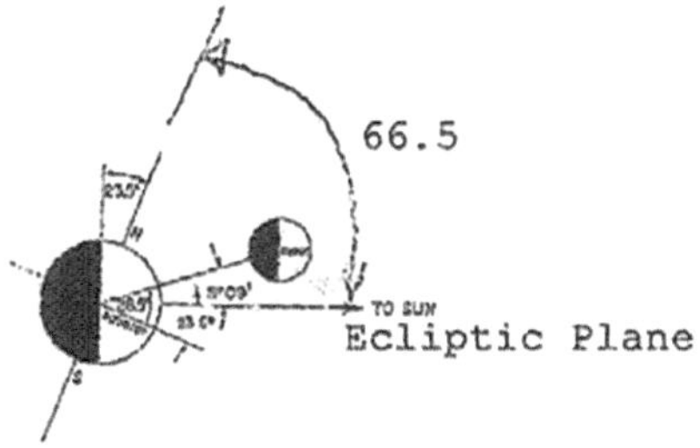

FIGURE 3.9 Interaction of sun, moon and earth.

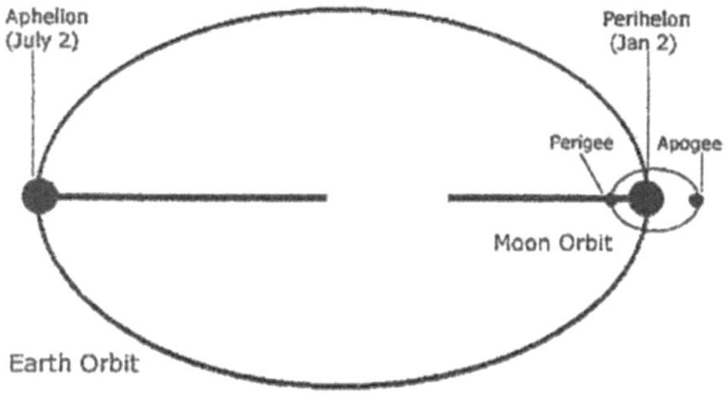

FIGURE 3.10 Interacting effects of the sun and the moon on the earth's tides.

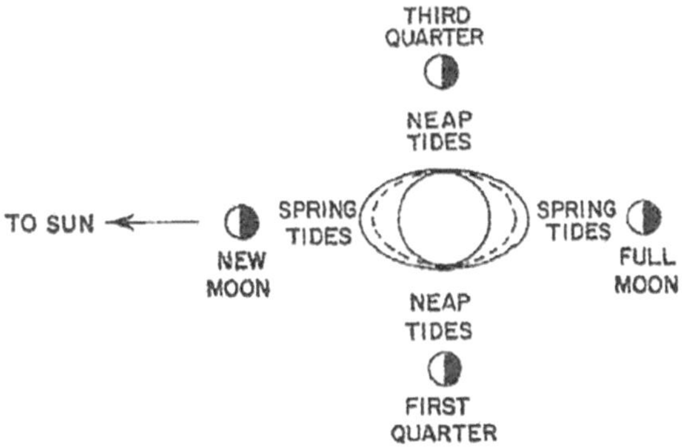

FIGURE 3.11 Sun and moon at spring and neap tides.

presented in Figure 3.10. A summary of the relative positions of the sun and moon at spring and neap tides is shown in Figure 3.11

3.4.2 Diurnal and Semi-Diurnal

The tide prediction problem is difficult because of the complex geometry of the oceans, shorelines and their basins. Semi-enclosed or enclosed basins tend to oscillate at some natural frequency or harmonic, so that when its natural period is close to the tide-producing period, the local tide will be amplified. Tides propagate everywhere at sea at the long-wave speed because the length of the tide wave is very long everywhere compared to the shallow ocean basins. At sea and at oceanic islands, the tidal amplitude is very small (approximately 1 ft or less). It is only in relatively shallow water such as found over the continental shelves that the amplitude begins to build.

The definitions of semi-diurnal (every 12 h), mixed, and diurnal (daily) tide plots are illustrated in Figure 3.12. Tidal theories are treated in detail in a number of physical oceanography texts and papers such as Dean (1966).

3.4.3 Tidal Currents

The design of marine structures requires the determination of a maximum water level and sometimes a minimum water level that will be experienced over the projected life of the project. These expected maximum and minimal water levels are termed the

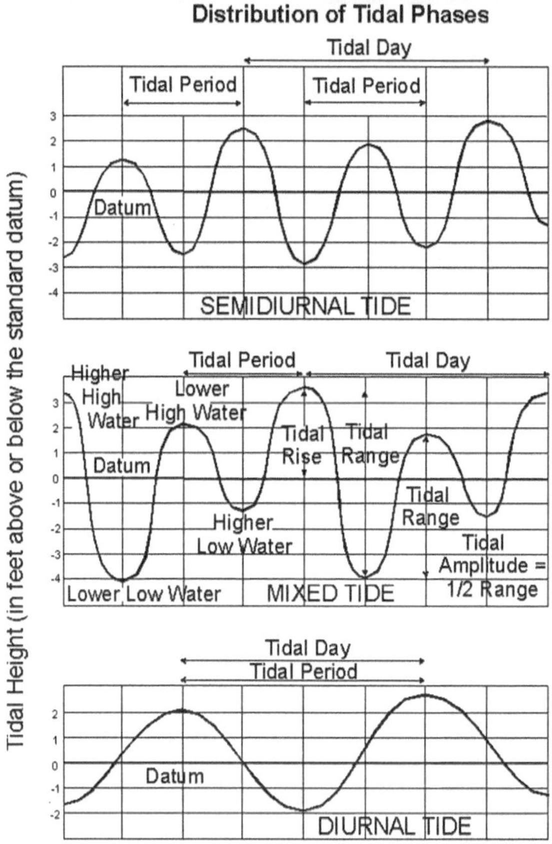

FIGURE 3.12 Our Restless Tides National Ocean Service (NOAA, http://co-ops.nos.noaa.gov/restles4.html).

Design Water Levels (DWL). The philosophy behind the requirement for an maximum DWL is that deck elevations must be above the maximum water elevation plus the crest elevation of the highest wave expected at the site. In addition, for structures such as jacket platforms it is usual practice to provide an air gap between the elevation of the crest of the highest wave and the underside of the platform. In contrast, a minimum DWL is sometimes required in the design of wharf or pile structures. In a wharf structure a decrease in the water elevation on one side will cause an increase in the lateral load that must be resisted. In a pile structure a decrease in the water level may expose the pile/soil contact to scour.

The total design water level (DW.) is the summation of the various vector components as shown in Equation 3.1.

$$\text{DWL} = d + A_s + W_s + P_s + W_w \tag{3.1}$$

where

- d is the nominal water reference depth, typically MLW
- A_s is the astronomical tide at the time of the surge
- W_s is the wind setup
- P_s is the pressure or barometric setup
- W_w is the wave setup

3.5 CURRENTS

The causes of ocean currents is wind and density differences between water masses. The various types of currents is presented in Figure 3.13. Winds sweeping across the ocean exert a frictional force on the sea surface. This force horizontally displaces the affected water mass resulting in circulation. The speed and direction of currents are modified by a number of factors. These include the Coriolis effect, the shape of ocean basins, and the presence of land masses.

Thermohaline circulation is the result of density differences between water masses. This results in the sinking of surface waters. Thermohaline circulation is thus responsible for a slow overturning of the oceans waters. This mixing recharges the surface waters with nutrients.

3.5.1 Surface Wind-Driven Currents

3.5.1.1 Wind-Driven Circulation

Oceanic phenomena depend upon the physical interaction between the atmosphere and the ocean (Figure 3.13). Winds in the lower atmosphere are responsible for waves, surface currents, mixing of the surface layer of the ocean, and the exchange of energy in the form of heat and water vapor. Particulate matter ranging from mineral grains, seeds, and spores to carbon and silica spherules are transported out to sea through atmospheric circulation.

Currents due to winds are set in motion by a combination of moving air masses and shaped by the earth's rotation and the configuration of the continents. Moving air affects the sea surface and the waters move in response. The relationship between

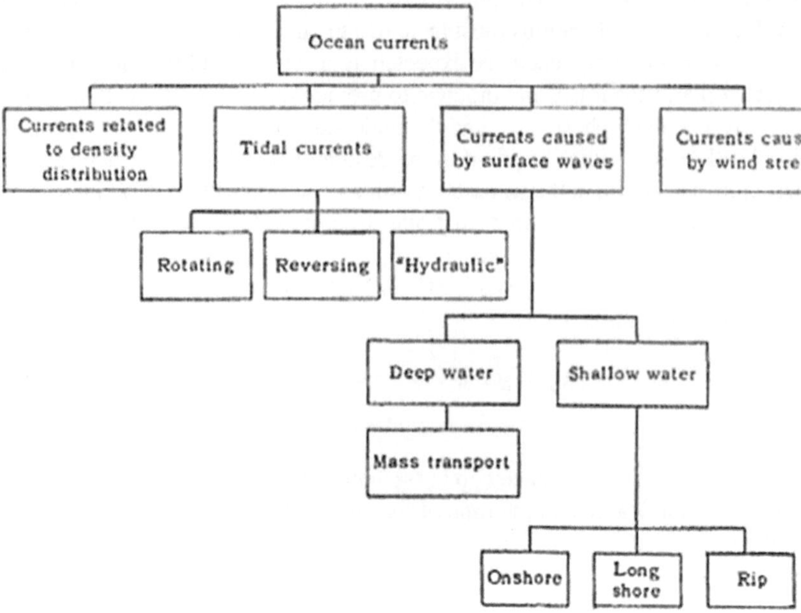

FIGURE 3.13 Types of ocean currents (Richards, 1981).

wind speed and water is not straightforward. Water masses in the upper layers flow at an angle to the wind direction due to the Coriolis effect. Assume initially a stationary sea surface in the northern hemisphere. A gentle breeze begins to blow, gradually intensifying over time. As the wind blows across the sea surface, a column of water, whose depth is a function of wind speed, duration, and fetch moves in response. The column of water does not move uniformly. As the wind exerts a frictional stress on the water surface layer, the water begins to move in the direction of the wind but is immediately acted upon by the Coriolis affect. This deflects the water motion to the right of the wind direction in the northern hemisphere and to the left in the southern hemisphere. Simultaneously, fictional stress between the surface layer and the water layer below it retards the movement of the surface layer. The resultant flow of the surface layer represents a steady state condition in which the surface water moves in a direction 45° to the right of the wind direction in the northern hemisphere or 45° to the left of the wind direction in the southern hemisphere.

The net movement of the water column affected by the wind can be visualized as a series of individual layers of water. By applying the same principles as those presented for the movement of the surface layer, it can be shown that each progressively deeper layer of water moves to the right of the layer above it and at a slightly lower velocity. The movement of the water layers as a series of vectors whose length corresponds to relative velocity is shown in Figure 3.14. A review of this figure shows that it resembles a spiral staircase with successively shorter steps arranged at progressively greater angles from the wind direction. This circulation pattern is called the Ekman spiral after Swedish physicist V.W. Ekman, who first described it mathematically. The effect of

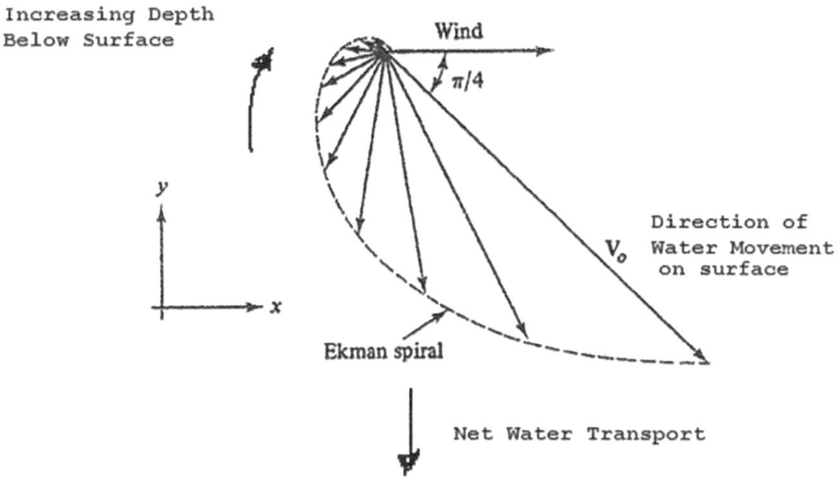

FIGURE 3.14 Net movement of water column affected by wind.

both the deflecting for the earth's rotation (Coriolis effect) and of the eddy viscosity (A) was taken into account by Ekman (1902). The eddy viscosity is defined as follows:

$$\tau_s = A\frac{dv}{dn} \tag{3.2}$$

where

$\frac{dv}{dn}$ is the shear of the observed velocities

A is the expression for the transfer of momentum of mean motion

τ_s is the shearing stress

Eddy viscosity, depends upon state of motion of the fluid and is not a characteristic physical property

Assuming the eddy viscosity is independent of depth (D) then the following expression can be written:

$$\sigma_x = C_1 e^{\frac{\pi}{D}z}\cos\left(\frac{\pi}{D}z + c_1\right) + C_2 e^{\frac{-\pi}{D}z}\cos\left(\frac{\pi}{D}z + c_2\right) \tag{3.3}$$

$$\sigma_y = C_1 e^{\frac{\pi}{D}z}\sin\left(\frac{\pi}{D}z + c_1\right) - C_2 e^{\frac{-\pi}{D}z}\sin\left(\frac{\pi}{D}z + c_2\right) \tag{3.4}$$

$$D = \pi\sqrt{\frac{A}{\rho\omega\sin\varphi}} \tag{3.5}$$

where C_1, C_2, c_1, c_2 are constants that depend on the boundary conditions.

Assuming D is a large number then the motion near the bottom of the water column is zero, then $C_1 = 0$. In addition, assume that the stress of the (τ_a) is along the y axis then

$$\tau_a = -A\left(\frac{dV_y}{dz}\right)_0 \quad (3.6)$$

$$0 = A\left(\frac{dV_x}{dz}\right)_0 \quad (3.7)$$

allows C_2 and c_2 to be determined. If V_0 is the velocity at the surface then

$$V_x = V_0 e^{\frac{-\pi}{D}z} \cos\left(45° - \frac{\pi}{D}z\right) \quad (3.8)$$

$$V_y = V_0 e^{\frac{-\pi}{D}z} \sin\left(45° - \frac{\pi}{D}z\right) \quad (3.9)$$

$$V_0 = \frac{\tau_a}{\sqrt{\rho A 2\omega \sin\varphi}} = \frac{\pi\tau_a}{D\rho\omega \sin\varphi\sqrt{2}} \quad (3.10)$$

Therefore the direction of water movement in the northern hemisphere is 45° to the right of the wind as shown schematically in Figure 3.14. In the southern hemisphere the water motion is to the left of the wind direction,

Under ideal conditions, the Ekman spiral results in a net transport (Ekman transport) of wind-driven water in the affected water column (Ekman Layer), which is 90° to the right from the wind direction in the northern hemisphere and to the left in the southern hemisphere. In shallow water it will be somewhat less than 90° because of the restricting influence of the ocean bottom.

3.5.1.2 The Oceanic Gyres and Geostrophic Currents

A characteristic feature of the major surface currents is that they describe large circular orbits, or gyres, within individual ocean basins. The circulation of these gyres is anticyclonic (i.e., clockwise in the northern hemisphere and counterclockwise in the southern). Typical of the ocean basin gyres is a major clockwise current flow in the North Pacific called the North Pacific gyre. The gyre itself resembles a hill with its peak within the gyre. The hill in oceanic gyres only rise about 2 m above the base of their slopes. Nonetheless, the hills have a significant effect on ocean circulation.

The mounding up of water in oceanic gyres is caused by Ekman transport. The net transport of an entire column of wind-driven water is approximately 90° from the wind direction and about 45° from the direction of the uppermost surface currents as discussed previously. Consequently, there is a continual transport of surface waters toward the interior of a gyre. This convergence of waters results in a slight rise of the sea surface. The mounding up of relatively warm, low-density surface waters in the interior of the gyres depresses the colder, higher density waters below.

Consider a water element on the slope of this mound. In response to gravity the element begins to move down the mound; however, once the element is in motion the Coriolis effect deflects the water element to the right of its downward path. The balance established between the pressure-gradient force and the Coriolis effect causes the surface waters in the gyre to follow a curved path conforming to the contours of the mound as shown in Figure 3.15 and limits the slope and height of the mound. The net effect is that the water element moves around the mound.

When the Coriolis effect and the pressure gradient force are in balance, they are said to be in geostrophic balance. The movement of surface waters around an oceanic gyre is therefore called geostrophic currents. The geostrophic currents are responsible for the circular movement of waters in oceanic gyres and conceivably would serve to maintain the gyres for extended period of time even if the global winds suddenly ceased.

3.5.1.3 Westward Intensification of Ocean Currents

The individual currents that comprise a specific oceanic gyre do not flow at the same speeds. Currents along the western margins of ocean basins are considerably swifter and narrower than their counterparts on the eastern side. An example of this effect is the velocity of the Gulf Stream approaches 9 km/h, while the Canary Current flows south along the European coast at less than 1 km/h (Parker, 1985). This western intensification of gyre currents results from a balance between several factors that affect the surface currents. These include the increasing magnitude of the Coriolis effect with latitude, latitudinal variations of wind strength and direction, and friction between the water currents and contiguous land masses. These factors cause the topographic highs of the oceanic highs of the oceanic gyres to be offset to the west as shown schematically in Figure 3.15. The western boundary currents move faster than those on the eastern side because a constant volume of water is squeezed into a narrower band along the western margins of the ocean basins.

3.5.1.4 Ekman Transport and Vertical Water Movement

Horizontal displacement winds cause substantial vertical displacement of water masses adjacent to continental margins. A phenomenon known as up-welling occurs

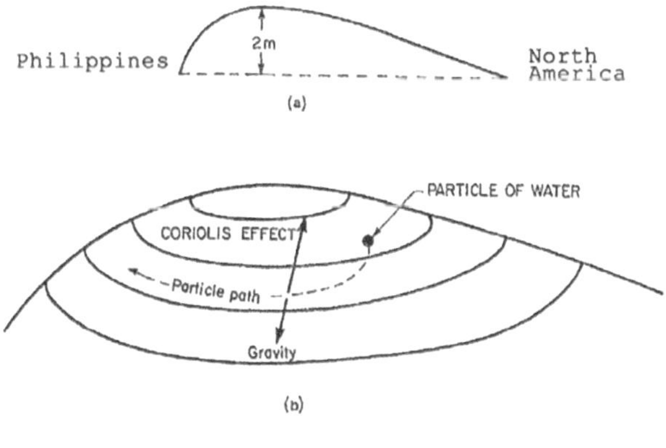

FIGURE 3.15 Coriolis effect on surface water in a gyre (Adapted from Parker, 1985).

when wind patterns cause a continuous mixing of water from depths to the surface. A reverse vertical flow called down welling can occur under other wind circumstances. Wind-induced vertical mixing has major effects on biologic productivity and the resulting occurrence of biogenic sediments. A schematic diagram illustrating the mixing phenomena occurring in a semi-enclosed marginal sea such as the SCS is shown in Figure 3.16. Figure 3.16 is divided into three cases of mixing based on water depth, slope of seabed, evaporation, precipitation, and runoff.

Upwelling occurs next to continental margins and along the equator when prevailing winds cause Ekman transport of surface water away from the affected area. Deep waters then move upward to replace the water elements. In the northern hemisphere this will occur when the coast is to the left of the wind direction and in the southern hemisphere when the shoreline is to the right of the prevailing wind (Figure 3.16). Deep, cold waters rich in the nutrients that stimulate growth of microscopic floating plants (i.e., phytoplankton) rise up to take the place of the vacating surface waters. Stocks of commercially valuable fish produce a highly productive ecosystem. Zones where this process occurs are along the coast of Peru, along the west coast of North American, and the east coast of Eurasia. Equatorial upwelling occurs along the equator, because of the Ekman transport of surface water away from the equator due to westward blowing trade. To take the place of the diverging water masses deep, cold water then rises up. When water is piled up along a coast due to Ekman transport toward shore down welling occurs.

3.5.1.5 Wind-Driven Circulation in the North Pacific

The North Pacific is an example of a large-scale wind driven circulation pattern. The circulation patterns of the major oceans characteristically involve both a large central gyre and one or more secondary gyres.

The North Pacific general circulation pattern of surface waters is deflected to the right and pushed toward the interior of the mid-Pacific gyre, where a high pressure zone is maintained. This is due to the trade winds that blow between 30° north latitude and the equator. This movement of surface waters establishes a major, westward flowing geostrophic current, the North Equatorial Current. The current reaches the western margin of the North Pacific Basin, where land masses deflect it towards the north. There it merges with a portion of the Kuroshio Current, which has been deflected to the north by the Philippines. The combined current moves along the east coast of the

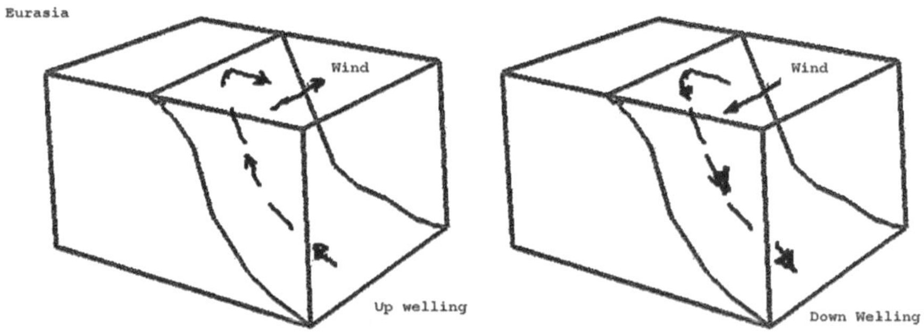

FIGURE 3.16 Coastal upwelling north and south of the equator.

Philippines, Taiwan, and Japan as the Kuroshio Current. The Kuroshio Current then merges with the North Pacific current, which in turn merges with the California current along the west coast of North America.

When the North Pacific Current reaches the western margin of the North Pacific Basin, part of it flows north and then west. Another branch moves into the Bering Sea as the Aleutian Current. The rest flows south along the coast of North America as the California Current. Eventually, the California Current rejoins the North Equatorial Current, completing the closed loop of the North Pacific gyre.

Between the North and South Equatorial Currents flows a narrow, eastward-flowing current called the Equatorial Counter-current. Other eastward flowing currents move along the equator in the Pacific, Atlantic, and Indian (seasonally) Oceans in addition to the equatorial countercurrents. These are subsurface currents, generically referred to as equatorial undercurrents. An example is the Pacific equatorial undercurrent, also known as the Cromwell Current.

Equatorial undercurrents are approximately 250 m thick and 250 km wide and reach speeds of up to 5 km/h. The causes of equatorial undercurrents are not completely understood. It has been suggested that the equatorial undercurrent is an eastward return flow of water resulting from the North and South Equatorial Currents. This flow would then be a result of the pressure gradient caused by the slope of the sea surface.

3.5.2 Vertical Currents, Density Driven

3.5.2.1 Thermohaline Circulation

The thermohaline circulation in the oceans is driven by the sinking of water masses, primarily in high latitudes, in response to temperature and salinity changes. Before describing the details of thermohaline circulation, it is important to first discuss the patterns of salinity, temperature, and density in the oceans.

Sea water is slightly denser than fresh water due to its salt content. Ignoring pressure effects, its density usually lies between 1.024 and 1.028 g/cm^3. The density of water is a function of its pressure, temperature, and salinity. The pressure effect results from reduction in volume attributable to compression. Since water is slightly compressible (a 1% reduction in volume occurs at a depth of 2 km), the enhancing effect of pressure on density is important only for deep water.

An increase in density can also result from an increase in salinity or a decrease in temperature. To determine the density of a specific water mass, plots of temperature (T) versus salinity (S) are constructed. The T-S diagram results in lines of constant density. The T-S diagram allows the reconstruction of the density profile of a column of water from profiles of salinity and temperature.

The density of seawater with depth depends on how well the oceans are mixed. This information can be determined based on correlations of temperature, salinity, and density with depth. Seawater density is expected to be uniform from top to bottom only if the oceans are thoroughly mixed in the oceans.

The oceans are highly stratified (i.e., poorly mixed) as demonstrated by decades of physical sampling of the ocean depths. The stratification is indicated by vertical profiles of temperature and salinity that reveal layers of water masses with sharply defined boundaries. The reason for large changes in ocean density occurring in the upper layers of the seas is because they are exposed to the energy of the sun and the dynamics of the atmosphere.

The atmospheric temperature and relative rates of evaporation and precipitation are reflected in the temperature and salinity of surface water. A layer of warm, low-density surface water overlies colder, denser water below when air temperatures are warm year-round and evaporation rates are relatively low. A zone of rapidly declining temperature (thermocline) and rapidly increasing density (pycnoline) exists between the two layers. These types of water columns are resistant (i.e., stable) to mixing. Stable water columns are characteristic of the topics. These types of tropical seas are frequently devoid of marine life due to the absence of mixing prevents the introduction of nutrients, trapped in deeper waters below.

Seasonal thermoclines exist in higher latitudes in the late fall because of an unstable water column. The instability is caused because, once cooled, the surface waters become denser than the underlying water. The surface water then sinks mixing with the lower less dense nutrient laden layer.

Some sinking of high density waters also occurs in lower latitudes. Here, the elevated density is strictly a function of high salinity because of high evaporation rates. The waters of the Mediterranean and the western Pacific provide an example. Because evaporation far exceeds precipitation and runoff from the surrounding land in the virtually enclosed Mediterranean and marginal ocean basins, the high salinity of surface waters causes continual sinking. As an example deep, saline (up to 39 parts per thousand) water flows out over the Gibraltar sill into the Atlantic Ocean. There it eventually reaches an equilibrium depth of about 1.2 km (the depth at which the Atlantic Ocean water has the same density). The Mediterranean water mass then spreads throughout the Atlantic at that depth, where it is detectable far from the Strait of Gibraltar. A similar process takes place in the South China Sea (refer to Figure 3.17).

3.5.3 Water Exchange with Oceans

Both river and oceanic currents enter the marginal seas where they are able to mix (Figure 3.17). The hydrologic system in the marginal sea is thus highly sensitive to sea-level fluctuation during glacial-interglacial cycles. Even a minor drop of the sea level can cause major changes in water circulation and consequently the nature of regional land–sea interation (Wang, 2004).

Water of Pacific Ocean origin that enters the SCS through the Bashi (Luzon) Strait is lower in temperature and higher in salinity, and from there, part of the water continues southward into the Java Sea and returns to the Pacific through the Makassar Strait. This circulation has been coined the "SCS throughflow" (Figure 3.17), acting hypothetically as a heat and freshwater conveyor in the regulating seawater salinity and temperature (SST) pattern in the SCS.

The main passageways (black dots) and rivers emptying into the marginal seas are shown in Figure 3.18. Western boundary currents flow through the marginal seas. Numbers give the water in Sv (1 Sv = 10^6 km/s).

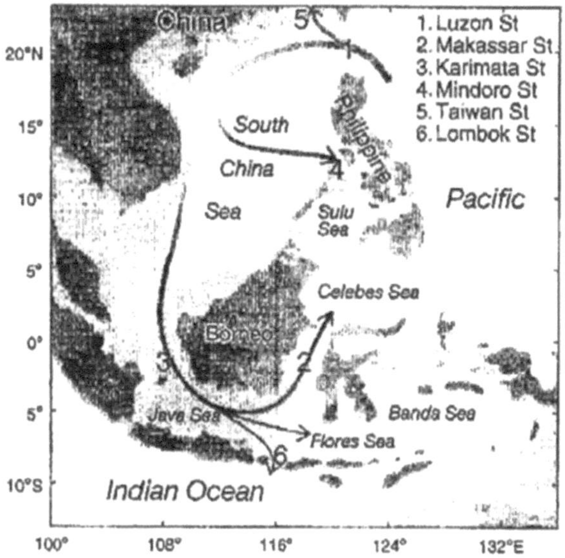

FIGURE 3.17 A schematic diagram of the South China Sea through flow (Qu et al., 2006). Reprinted with permission of Elsevier.

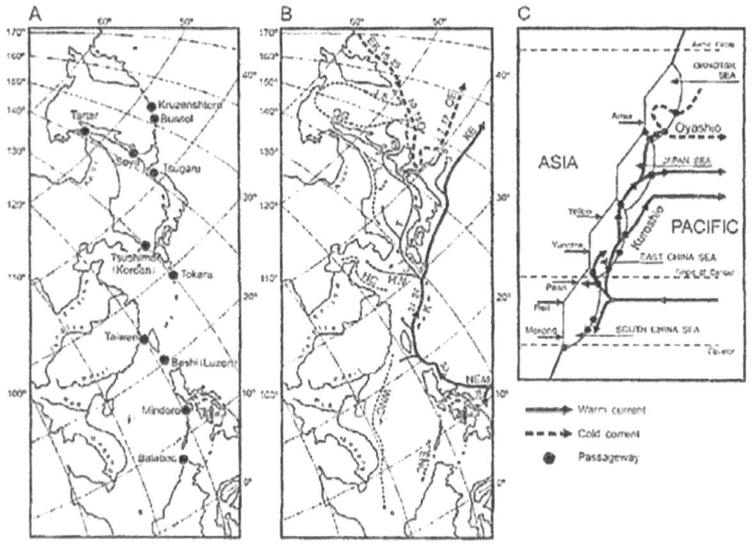

FIGURE 3.18 A simplified sketch showing the hydrographic system in the western Pacific marginal sea between the Asian continent and the Pacific Ocean (Okhotsk, Japan, and East China and South China Seas) and the western boundary currents: EK East Kamchatka Current; NNC, East Nansha Coast Current; HC, Huanghai (Yellow Sea) Coastal Current; Huanghai Warm Current; K, Kuroshio Current; KE, Kuroshio Extension; NE, North Equatorial Current; O, Oyashio Current; OE, Oyshio extension; OG, Okhotsk, Gyre; S, Soya Current; T, Tsushima Current (Wang, 2004). Reprinted with permission of Elsevier.

REFERENCES

Dean, R.G. (1966). "Tides and Harmonic Analysis," in A.T. Ippen (Ed.), *Esturary and Coastline Hydrodynamics*, McGraw-Hill, New York, pp. 197–230.

Fox, W.T. (1983). *At the Sea's Edge*, Prentice-Hall, Englewood Cliffs, NJ.

Parker, H.S. (1985). *Exploring the Oceans: An Introduction for the Traveler and Amateur Naturalist*, Prentice-Hall, Englewood Cliffs, NJ, 354pp.

Qu, T., Girton, J.B., and Whitehead, I.A. (2006). "Deepwater overflow through Luzon Strait," *Journal of Geophysical Research*, 111: C01002. http:/dx.doi.org/10.1029/2005JC003139.

Qu, T., Song, T., and Yamagata (2009). "An introduction to the South China Sea throughflow: Its dynamics, variability, and application for climate," *Dynamics of Atmospheres and Ocean*, 47(1): 3–14.

Richards, A.F. (1981). Personal communication.

Wang, P. (2004). "Cenozoic deformation and the history of sea-land interactions in Asia," in P. Clift, P. Wang, D. Hayes, and W. Kuhnt (Eds.), AGU Geophysical Monograph: Vol. 149, *Continent Ocean Interactions in the East Asian Marginal Seas*, American Geophysical Union, Washington, DC, pp. 1–22.

Wiegel, R.L. (2013). *Oceanographic Engineering*, Dover, New York.

4 Wave Characteristics

4.1 INTRODUCTION

Ocean surface waves refer to a moving succession of irregular crests and troughs on the ocean surface. A slight disturbance of the sea surface will be exhibited as tiny round ripples or capillary waves as the wind begins to blow (Figure 4.1a). Capillary waves have very small wavelengths (λ), short periods (T) and small wave heights (H). Returning the sea surface to an undisturbed glassy state requires damping of these capillary waves. The damping force is surface tension. This damping or restoring force results in the propagation of the wave in a horizontal direction. This is similar to a small wave traveling along a tightly stretched piece of fabric.

Waves grow in length, height, and energy as the wind continues to blow. The growth of waves results because the disturbed sea surface is increasingly more directly exposed to the wind. When the wave lengths exceed 1.74 cm, gravity replaces surface tension as the dominant restoring force. The resultant wave form is called a gravity wave (Figure 4.1b). A short choppy sea characteristically then develops because of the interactions between different waves creating a variety of wave forms with different wavelengths. Wave height increases more rapidly than wavelength as more energy is imparted to the waves by the wind. This process continues until the resultant wave steepness causes them to break (Figure 4.1c). In breaking waves, or whitecaps, the energy received from the wind is balanced by the energy lost in breaking.

The same fundamental laws of physics controlling sound and light waves also apply to ocean waves. Waves represent the propagation of mechanical energy due to wind, earthquakes, volcanic activity, landslides, meteorological phenomena, or even by other waves. An important property of waves under ideal conditions is that there is no net displacement of the particles set in motion by a wave. Water molecules

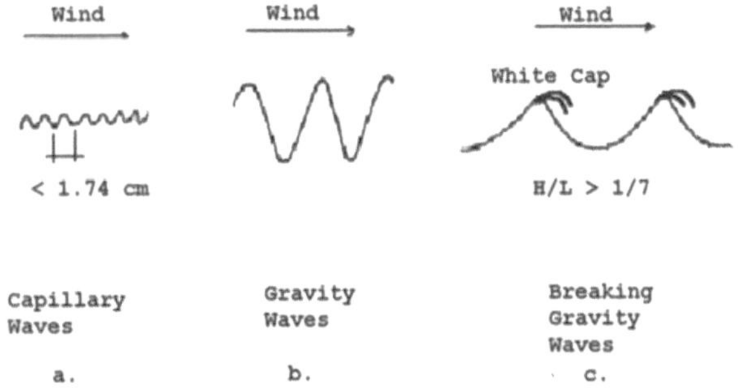

FIGURE 4.1 Development of waves on an initially smooth sea surface in response to a sustained unidirectional wind.

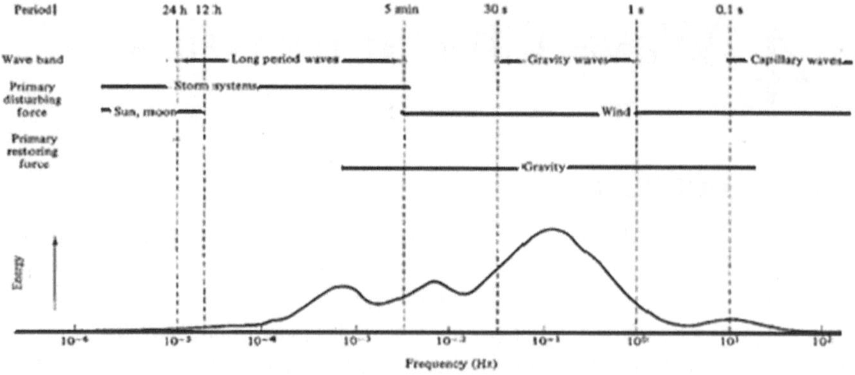

FIGURE 4.2 Classification of surface waves and schematic representation of their power spectrum (Adapted from Kinsman, 1965).

affected by an ocean wave describe a circular orbit but undergo only a slight net forward displacement. The orbital motion of particles is characteristic of waves occurring between fluids of different densities. Waves move along the air/water interface but also along any pycnoline. There is a complete spectrum of waves ranging from small capillary waves (2 cm long) to the tides with wavelengths of thousands of kilometers. The properties of ocean surface waves and a schematic representation of their power spectrum is presented in Figure 4.2.

4.2 GRAVITY WAVE GENERATION

Three wave theories have been developed to explain the relationship between surface-wave data and water velocity, acceleration, and pressure beneath the waves for engineering purposes. These theories are the following: (1) linear (Airy) wave theory, (2) non-linear Stokes wave theory, and (3) cnoidal wave theory.

4.2.1 Linear Wave Theory (Airy Theory)

A simple theory of wave motion, known as the Airy wave theory, was developed by G.B. Airy in 1842. This theory assumes a sinusoidal wave form whose height H is small in comparison with the wave length λ and the water depth h (Figure 4.3). The theory is useful for preliminary calculations and for illustrating the basic characteristics of wave-induced water motion. It also serves as a basis for the statistical representation of waves and the induced water motion during storm conditions.

A surface wave amplitude (η) of point A as shown in Figure 4.3 as a function of time can be described by the following equation.

$$\eta = \frac{H}{2}\cos\left(kx - \omega t + \psi\right) \tag{4.1}$$

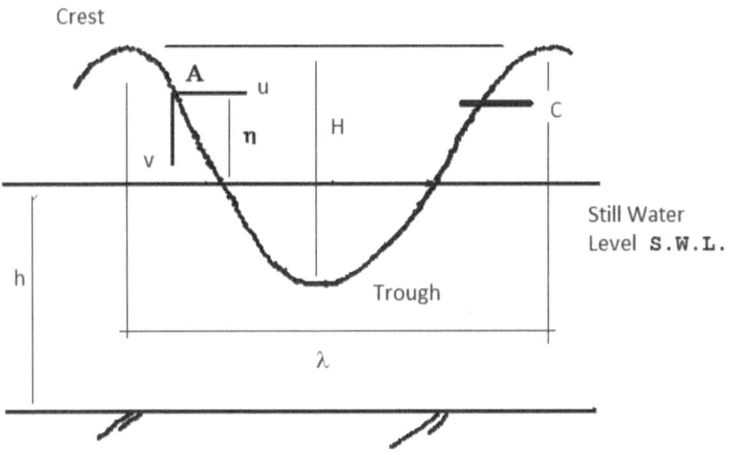

FIGURE 4.3 Definition of wave properties.

The corresponding horizontal velocity (u) and vertical velocity (v) of the water particle at point A (x, y) and time t are expressible according to the appropriate hydrodynamic equations. These relationships have been shown by a number of authors including Kinsman (1965) and McCormick (1973).

$$u = \frac{wH}{2} \frac{\cosh\ ky}{\sinh\ kh} \cos\left(kx - wt + \psi\right) \tag{4.2}$$

$$v = \frac{wH}{2} \frac{\sin\ hky}{\sin\ hkh} \sin\left(kx - wt + \psi\right) \tag{4.3}$$

where

- k is the wave number = $2\pi/\lambda$
- ω is the angular radian frequency = $2\pi/T$
- T is the wave period
- λ is the wave length
- H is the wave amplitude
- X is the horizontal direction
- Ψ is the phase angle

These terms are related to each other by the following equation, which expresses the circular frequency of the wave as shown below. The relationship between radian frequency and wave number depends on water depth and is given in Equation 4.4.

$$\omega = \left[gk \tanh kh\right]^{1/2} \tag{4.4}$$

where

- g is the acceleration of gravity
- h is the water depth

This relationship is called the wave dispersion equation. For large values of kh, Equation 4.4 becomes

$$\omega^2 = gk$$

If the term $(kx - \omega t)$ in Equations 4.2 and 4.3 remain unchanged at time $(t + \Delta t)$ as shown in Figure 4.4 then the wave amplitude (H) remains the same.

This can occur if Δx equals the following:

$$\Delta x = (\omega / k)\Delta t \tag{4.5}$$

then

$$kx - \omega t = k(x + \Delta x) - w(t + \Delta t) \tag{4.6}$$

The surface wave described by Equation 4.1 can therefore easily be seen to represent a fixed wave form propagating to the right with a speed or celerity c. The celerity (c) can be related to wavelength (λ) and period (T) using Equation 4.7.

$$c = \frac{\lambda}{T} = \frac{w}{k} \tag{4.7}$$

Substituting Equation 4.4 into Equation 4.7 gives the following relationship.

$$c = \left(\frac{g}{k} \tanh kh\right)^{1/2} \tag{4.8}$$

Using Equation 4.8, harmonic waves can be divided into shallow, intermediate, and deep water waves depending on the ratio of water depth to wave length. This ratio also leads to three phase velocity classifications.

The above expressions for deep water and for shallow water can be simplified.

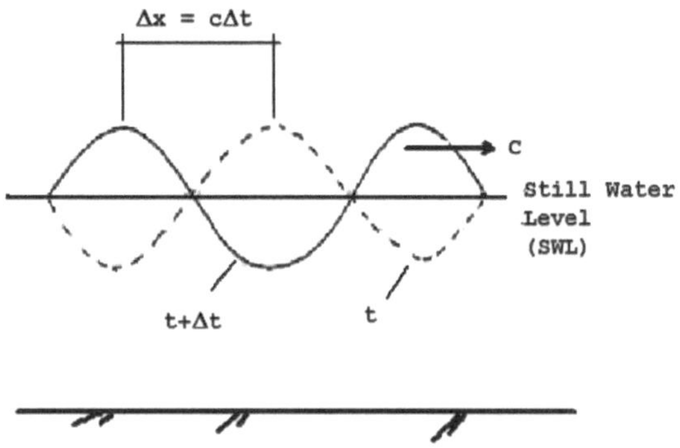

FIGURE 4.4 Propagating waves.

4.2.1.1 For Deep Water kh > π, 1/2 < h/λ < ∞

For this range of h/λ indicated

$$\tanh kh \sim 1$$

$$c = \left(\frac{g}{k}\right)^{1/2} = \left(\frac{\lambda g}{2\pi}\right)^{1/2} \tag{4.9}$$

where

g is the acceleration due to gravity

k is the wave number = $\frac{2p}{l}$

A review of Equation 4.9 indicates that the celerity of a wave (speed) in deep water depends upon its wave length (λ). Waves with longer wavelengths travel faster than waves with shorter wavelengths. At a site a distance from a source of wave generation in deep water long wavelength waves tend to arrive first followed by waves of shorter wavelengths. This spreading out of waves is called dispersion.

Combining Equation 4.7 with Equation 4.9 gives the following:

$$c = \frac{Tg}{2\pi} \tag{4.10}$$

or

$$\lambda = \frac{T^2 g}{2\pi} \tag{4.11}$$

Note that Equation 6.9 shows that $c\alpha\sqrt{\lambda}$

The speed of a swell depends upon the square root of its wavelength. Therefore in deep water, longer waves move faster than short waves as they propagate from a storm area. This process is causes dispersion and explains why the first waves to reach a distinct beach from a large storm are the longer waves.

4.2.1.2 For Shallow Water, kh < $\pi/10$ or h/λ < 1/20

$$\tanh kh \sim kh = \frac{2\pi h}{\lambda}$$

$$c = \left(\frac{2\pi gh}{k\lambda}\right)^{1/2} = (gh)^{1/2} \tag{4.12}$$

Thus the speed of a shallow water wave is the following

$$C = \sqrt{gh} \tag{4.13}$$

where h is the depth.

The gage pressure p (difference between actual pressure and atmospheric pressure) at any place (x, y) and time t resulting from the overhead wave and from the hydrostatic contribution can be determined using the following expression.

$$u = \rho_w g \frac{H}{2} \frac{\cosh ky}{\cosh kh} \cos(kx - wt) + \rho_w g(h - y) \tag{4.14}$$

where ρ_w is the mass density of water.

4.2.2 Stokes Wave Theory

An extension of the Airy theory to waves of finite height was made by G.G. Stokes in 1846. His method was to expand the wave solution in series form and determine the coefficients of the individual appropriate hydrodynamic equations for finite-amplitude waves.

Stokes wave theory (second, third, or fifth order) is valid for non-linear waves on intermediate and deep water. That is, for wave lengths (λ) that are not large compared to water depth (h). In shallow water, the low order Stokes expansion breaks down. That is, it tends to give unrealistic results.

Stokes carried his analysis forward to third order of accuracy in the wave steepness H/λ. This solution has been presented by Skjelbreia (1969) and Wiegel (2013). An extension of the method to fifth order has been made by Skjelbreia and Hendrickson (1961). This work is typically referred to as Stokes fifth-order wave theory and is used for finite amplitude waves.

Due to numerical difficulties the theory is considered valid for water where the relative depth h/λ is greater than 1/6.

For a wave of height H, wave number k, and frequency w propagating in the positive x-direction, the free-surface water deflection η from the still-water level is, according to the Stokes fifth-order theory, given by Equation 4.15.

$$\eta = \frac{1}{k} \sum_{n=1}^{5} Fn\cos n(kx - wt) \tag{4.15}$$

where F_1, F_2, F_3, etc. are given by the following:

$$\begin{aligned} F_1 &= a \\ F_3 &= a^2 F_{22} + a^4 F_{24} \\ F_3 &= a^3 F_{33} + a^5 F_{35} \\ F_4 &= a^4 F_{44} \\ F_5 &= a^5 F_{55} \end{aligned} \tag{4.16}$$

with F_{22}, F_{24}, etc. denoting wave-profile parameters dependent on kH and parameter a denoting a wave height parameter related to the wave height through Equation 4.17.

$$\mathrm{kH} = 2\left(a + a^3 F_{33} + a^5\left(F_{35} + F_{55}\right)\right) \tag{4.17}$$

The horizontal water velocity u and the vertical water velocity v (at place x, time t, and distance y above the seafloor) caused by the free-surface wave propagating over water of depth h are expressible as

$$u = \frac{w}{k}\sum_{n=1}^{5} Gn\frac{\cosh nky}{\sinh nkh}\cos n\left(kx - wt\right) \tag{4.18}$$

$$v = \frac{w}{k}\sum_{n=1}^{5} Gn\frac{\sinh nky}{\sinh nkh}\sin n\left(kx - wt\right) \tag{4.19}$$

where

$G_1 = aG_{11} + a^3G_{13} + a^5G_{15}$
$G_2 = 2(a^2G_{22} + a^4G_{24})$
$G_3 = 3(a^3G_{33} + a^5G_{35})$
$G_4 = 4a^4G_{44}$
$G_5 = 5a^5G_{55}$
G_{11}, G_{13}, etc. denote wave velocity parameters dependent on kh.

Explicit expressions for F_{22}, F_{24}, G_{11}, etc. are given by Skjelbreia and Hendrickson (1961).

In addition to the above it is also necessary to have the frequency relation connecting wave frequency with wave number.

$$w = gk\left(1 + a^2{}_1 + a^4C_2\right)\tanh kh \tag{4.20}$$

where C_1 and C_2 are frequency parameters.

The wave speed C is determined as in the Airy theory from the relation $C = w/k$, which for the Stokes fifth-order solution is expressible as given below.

$$C = \left(\frac{g}{k}\left(1 + a^2C_1 + a^4C_2\right)\tan hkh\right)^{\frac{1}{2}} \tag{4.21}$$

The gage pressure (p) in the water resulting from the overhead wave and the hydrostatic contribution can also be determined from the velocity components by substitution into the following equation

$$p = \rho_w\frac{w}{k}u - \frac{1}{2}r\left(u^2 + v^2\right) = \frac{\rho_w g}{k}\left(a^2C_3 + a^4C_4 + ky'\right) \tag{4.22}$$

where $y' = y - h$ and where C_3 and C_4 denote pressure parameters dependent on kh or h/λ.

4.2.3 Cnoidal Wave Theory

For shallow water, the cnoidal wave theory is generally regarded as being more correct than the Stokes wave theory. This theory was first presented by Korteweg and de Vires in 1895 and has since been developed further by several writers. Wiegel (2013) has summarized these developments and has given a presentation of the theory convenient for practical application.

Cnoidal waves are periodic waves with surface profiles (h) and are described in terms of the wave number k and frequency ω as given by Equation 4.23.

$$h = h_T + Hcn^2\left(kx - \omega t, m\right) \tag{4.23}$$

4.2.3.1 Fetch and Seas

As the wind begins to blow over the surface of the ocean an amount of energy is imparted to form wind waves. The first waves to form are capillary waves. These waves, in turn, make the sea somewhat rougher and allow more efficient interaction between the wind and the sea surface and therefore allows a more efficient transfer of energy from the wind to the sea.

The harder the wind blows, the greater the amount of energy transfer, the larger the waves. The size of the resulting waves (i.e., height and wavelength) is a function of the wind force, its duration, and the distance over which it blows (i.e., fetch). The minimum fetch and duration required for full development of waves associated with various wind speeds is present in Table 4.1.

A fully developed wave is the limiting condition. The characteristics of fully developed waves is presented in Table 4.2. The sea surface away from the storm has a smoother appearance. Waves of different lengths disperse in such a way that the longer ones increase in height.

TABLE 4.1
Full Wave Development Requirements

Wind Speed m/s (kt)	Fetch (km)	Duration (h)
6.1 (10)	16.5	2.4
6.2 (20)	140.	6.
16.3 (30)	520.	23.
20.4 (40)	1320.	42.
26.5 (50)	2570.	66.

Source: Data from Wolfe et al. (1966).

TABLE 4.2
Characteristics of Fully Developed Wind Waves

Wind Speed m/s (kt)	Average Period (sec)	Average Length (m)	Average Height (m)	Maximum Height (m)	Approximate Celerity (m/s) (kt)
6.1 (10)	2.9	6.5	0.27	0.55	6.6 (9)
6.2 (20)	6.7	32.9	1.5	3.0	6.7 (17)
16.3 (30)	6.6	76.5	6.1	6.5	13.3 (26)
20.4 (40)	6.4	136.0	6.5	16.3	16.8 (35)
26.5 (50)	16.3	212.0	16.8	30.0	21.9 (43)

Source: Data from Wolfe et al. (1966).

4.3 WAVE HEIGHTS IN DEEP WATER

The size of a wave (i.e., its height and wavelength) in deep water is determined not only by the force of the wind, but also by its duration and by the distance over which it blows, or its fetch. The Significant Wave Height (SWH or Hs) is defined traditionally as the mean wave height (trough to crest) of the highest third of the waves (H1/3).

4.4 SEA STATE

A sea state in oceanography is the condition of the free surface on a large body of water with respect to wind waves and swell at a specific location and time. A sea state is characterized by wave height, period, and power spectrum. The sea state varies with time, as the wind conditions or swell conditions change. The sea state can be determined by an experienced observer, or through instruments such as weather buoys, wave radar or remote sensing satellites. Because of the large number of variables involved in determining the sea state it cannot be quickly or easily determined. Therefore simpler scales are normally used to give an approximate but accurate description of sea conditions. The most frequently used scale is the World Meteorological Organization (WMO) sea state code (refer to Table 4.3 for the WMO sea state code). In addition, the character of sea swells can be described using Table 4.4. The WMO sea state code largely adopts the "wind sea" definition of the Douglas Sea State.

4.5 EARTHQUAKES AND TSUNAMIS

Tsunamis or seismic sea waves are impulsively generated, dispersive waves of relatively long period and low amplitude as observed for a number of sites as a result of the Chile earthquake of May 22, 1960. These waves are typically generated by sudden large-scale sea floor movements usually associated with severe, shallow focus

TABLE 4.3
World Meteorological Organization Sea State Code

WMO Sea State Code	Wave Height (m)	Characteristics
0	0	Calm (glassy)
1	0–0.1	Calm (rippled)
2	0.1–0.5	Smooth (wavelets)
3	0.5–1.25	Slight
4	1.25–2.5	Moderate
5	2.5–4	Rough
6	4–6	Very rough
7	6–9	High
8	9–4	Very high
9	Over 14	Phenomenal

TABLE 4.4
Character of the Sea Swell

Low	0 None
	1. Short or average
	2. Long
Moderate	3. Short
	6. Average
	6. Long
High	6. Short
	6. Average
	6. Long
	6. Confused

Note: Direction from which swell is coming should be indicated.

earthquakes. Usually an earthquake of at least 6.5–7.0 Richter magnitude and with focal depths of less than 30–40 miles is required to initiate such sea floor movements. Tsunamis may also be generated by underwater landslides, volcanoes, or explosions. An example of a large-scale sea floor movement is a simple fault in which tension in the basement rock is relieved by the abrupt rupturing of the rock along an inclined plane. When such a fault occurs, a large mass of rock and sediment drops rapidly and the support is consequently removed from a column of water that extends to the surface. The water surface oscillates up and down as it seeks to return to mean sea level, and a series of waves are produced. In contrast, a tsunami can also be generated if the basement rock fails in compression, the mass of rock on one side rides up and over that of the other, and a column of water is lifted. Another mechanism is a landslide or mass movement that is initiated by an earthquake. If the slide begins above the water, abruptly dumping a mass of rock and soil into the sea, waves are generated. If the slide occurs well below the surface of the sea, it also can create waves. Tsunamis can be highly destructive waves, especially at certain locations prone to tsunami run-up. Although they are almost undetectable at sea, because of their long wavelengths, with periods of a few minutes to an hour or more, and heights of only 1 or 2 ft or less, when they approach shallow water, shoaling, refraction, and possible resonant effects can cause run-ups of from several meters to upwards of approximately 30 m or more, depending on the tsunamis characteristics and the local typography. Tsunamis are often observed as a series of highly periodic surges that may continue over a period of several hours.

Since tsunamis are associated with seismic activity, the most destructive cases have been recorded in the North Pacific Ocean rim of fire. Susceptible areas of engineering importance are typically associated with low laying coastal regions. Historical records of tsunamis damage have been discussed by Bascom (1980) and there have been a number engineering investigations of tsunamis (Magoon 1965, Matlock et al. 1962, Murty 1977, Wilson and Torum 1968). The height characteristics of some recent tsunamis are shown in Table 4.5. Tsunamis are categorized as long waves (wavelength of 100 miles or more), therefore, travel times can be estimated by using either maps or for deep water a simple mathematical relationship can be utilized. Maps do not provide information on the height or the strength of the wave, only the arrival times. Due to the large wavelength tsunamis can travel everywhere in the sea at approximately the shallow water wave speed as given by Equation 4.24.

$$C = \sqrt{gd} \tag{4.24}$$

where

C is the wave celerity or velocity
g is the acceleration due to gravity
d is the water depth

Assuming an average depth for the Pacific Ocean of 12,000 ft, then a tsunami wave will travel at approximately 400 knots (knots × 1.8532 = km/h). Then if the location of the epicenter of the earthquake is known, the travel time of a tsunami between points can be estimated using nautical charts by summing the incremental travel times along a wave front. The time it takes a tsunami to travel a given distance is

TABLE 4.5
Highest Recent Record Tsunamis

Year	Height (m)	Location
1958	524	Lituya Bay
1963	250	Vajont Dam
1980	260	Spirit Lake
1964	70	Alaska Earthquake

estimated using Huygen's principle. This principle states that all points on a wave front are point sources for secondary spherical waves. Minimum travel times are computed over a grid starting at the earthquake epicenter. Times are then computed to all surrounding grid points from the starting point. The grid point with minimum time is then taken as the next starting point and times are computed from there to all surrounding points. The starting point is continually moved to the point with minimum total travel time until all grid points have been evaluated. This technique is discussed by Shokin et al. (1986).

There are a number of situations in which the estimated arrival times may not agree with the observed arrival times of the tsunami waves. This is due in part to the following:

- Bathymetry is not accurate in the vicinity of the epicenter
- Epicenter is not well located, or its origin time is uncertain
- Epicenter is on land and a pseudo epicenter off the coast must be selected
- Bathymetry is not accurate in the vicinity of the reporting station
- Nonlinear propagation effects may be important in shallow water
- Observed travel times do not represent the first wave but instead are later arrivals.

The damage due to a tsunami is normally from large hydrostatic and hydrodynamic forces along with the impact of water borne objects, overtopping with subsequent flooding and erosion caused by the high water velocities. The characteristics of tsunami of interest to engineers are primarily those associated with the nearshore environment. These characteristics are (1) run-up heights, (2) surge or bore velocities, and (3) return period.

4.6 STORM SURGES

A storm over near shore waters can generate large water level fluctuations if the storm is sufficiently strong and the near shore region is shallow over a large enough area. This is commonly known as a storm surge or meteorological tide. Storm activity can cause both a set up (rise) and set down (fall) of the water level at different locations and times, with the set-up predominating in magnitude, duration, and areal extent. Specific causes of water level change include the following: (1) surface wind stress, (2) Coriolis acceleration, (3) long-wave generation by a moving pressure disturbance, (4) atmospheric pressure differentials, and (5) precipitation and surface runoff.

Storm surge calculations require knowledge of the spatial and temporal distribution of wind speed, wind direction, and surface air pressure for the design storm conditions.

4.6.1 Wind Fastest Mile

The greatest wind speed to be expected at a particular site can be estimated from analysis of local daily weather reports. Due to the fluctuation of wind speed, records are averaged over the time required for a horizontal column of air 1 mile long to pass the measuring station. The fastest mile of wind is then defined as the highest wind speed measured in a single day, and the annual extreme fastest mile of wind is the largest of the daily maximums recorded during a single year.

Statistical projections of the gust wind speeds 30 ft above the earth that can be expected are shown in Table 4.6.

4.6.2 Gust Velocity

The strength of gusts depends upon the temperature gradient, the mean wind speed, and altitude. Gusts can occur in both horizontal and vertical directions. At heights below 100 ft the velocity of horizontal gusts is larger than vertical gusts. At heights of 100 ft or more above the surface, vertical gusts can be of the same order of magnitude as horizontal gusts. Gust factors are defined as the ratio of gust speed at 30 ft above the surface to the 1 min average wind speed. These factors have been proposed by Bretschneider (1969) and Gentry (1953) and are summarized in Table 4.6. Gust factors increase with elevation above the surface of the sea as given in Equation 4.25.

$$\frac{G_z}{G_{30}} = \left(\frac{Z}{30}\right)^{\frac{1}{12}} \tag{4.25}$$

where

G_z is the gust speed at elevation z

z is the elevation Z above surface of water

G_{30} is the gust speed at 30 ft above the surface of the water

Wind speeds in general can be corrected to any duration using the method presented by Vellozzi and Cohen (1968).

TABLE 4.6
Gust Speeds for Mean Hourly Speeds between 20 and 80 Knots

Duration of Gust	G_{30}/V_1	Reference
1.0 min	1.25	Bretschneider (1969)
5 s	1.48	Bretschneider (1969)
0.5 s	1.61	Bretschneider (1969)
5 s	1.2	Gentry (1953)
0.5 s	1.3	Gentry (1953)

Notes:

(1) G_{30} is the gust speed at 30 ft above the surface.
(2) V_1 is 1 min average wind speed.

REFERENCES

Bascom, W. (1980). *Waves and Beaches*, Bantam Doubleday Dell, New York.

Bretschneider, C.L. (1969). "Overwater wind and wind forces," in J.J. Meyers (ed.), *Handbook of Ocean and Underwater Engineering*, McGraw-Hill, New York, pp. 12–13.

Gentry, R.C. (1953). "Wind velocities during hurricanes," *Transactions of ASCE*, 120(2731): 169–180.

Kinsman. (1965). *Wind Waves*, Prentice Hall, Inc., Englewood Cliffs, NJ.

Magoon, O.T. (1965). Structural damage by tsunamis, in *Coastal Engineering Specialty Conference*, Santa Barbara, CA, October 1965, American Society of Civil Engineers, pp. 35–68.

Matlock, H., Reese, L., and Matlock, R.R. (March 1962). *Analysis of structural damage from the 1960 tsunami at Hilo, Hawaii.* Technical Report Defense Atomic Support Agency 1268, Structural Mechanics Research Lab, University of Texas, Austin, 95pp.

McCormick, M.E. (1973). *Ocean Engineering Wave Mechanics (Ocean Engineering)*, Wiley, New York.

Murty, T.S. (1977). Seismic sea waves Tsunamis. *Bulletin of the Fisheries Research Board of Canada*, Toronto, Canada, 337pp.

Skjelbreia, L. (1969). "Stokes' third order approximation: Tables of functions," Council on Wave research, Engineering Foundation, 337pp.

Skjelbreia, L., and Hendrickson, J. (1961). "Fifth order gravity wave theory," *Coastal Engineering Proceeding*, 1(7): 184–196.

Shokin et al. (1986). "Calculations of tsunami travel time charts in the Pacific Ocean," *Science of Tsunami Hazards*, 5: 85–113.

Vellozzi, J., and Cohen, E. (1968). "Gust response factors," *Proceedings of the Structural Division ASCE*, 94(ST-6): 1295–1313.

Wiegel, R.L. (2013). *Oceanographic Engineering*. Dover, New York.

Wilson, B.W., and Torum, A. [1966] (1968). *The Alaskan Tsunami of March 1964: Engineering Evaluation Technical Memorandum No. 25*. Coastal Engineering Research Center, Corps Engineers, U.S. Army, Washington, DC.

Wolfe, J.H., Silva, R.W., and Myers, M.A. (1966). "Observations of the solar wind during flight of IMP-1," *Journal of Geophysical Research*, 71: 1319–1340.

Part IV

Terrestrial and Seabed Sediments

5 Characterization of Marine Sediments

5.1 INTRODUCTION

Soil is any unconsolidated material composed of discrete solid particles and interstitial gas and/or pore liquids (Sowers and Sowers, 1961). The terms *sediment* and *soil* are used interchangeably, depending on whether the discussion is about engineering or geological. As products of various geological processes, soils are extremely variable mixtures of weathered minerals (i.e., naturally occurring, solid, chemical elements, or compounds formed by a geological process). These minerals are the result of fragmented rocks (i.e., aggregates of minerals or natural glasses), highly variable in composition, grain size, and spatial arrangement of their particles, which may contain small amounts of organic matter, mostly decaying plant debris and animal residues. The classification of sedimentary rocks that break down into soil is presented in Table 5.1. The pore fluids may chemically affect the nature of the soil. This effect is especially predominate if the soil is fine-grained and depends on its constituents, such as dissolved salts and organic compounds. Marine soils (i.e., sediments) are usually totally saturated (i.e., the interparticle voids are completely filled with seawater). In some cases, the interparticle voids can contain gases, mostly produced by biogenic degradation of organic matter or introduced through geochemical processes.

TABLE 5.1
Classification of Sedimentary Rocks

	Clastic	Clastic
Particle Size	Sediment	Sedimentary Rock
More than 2 mm	Boulder, cobble, pebble	Conglomerate
2 mm–0.063 mm	Sand	Sandstone
0.036 mm–0.008 mm	Silt	Shale or mudstone
Less than 1/126 mm	Clay	
	Non Clastic	Non Clastic
Siliceous	Limey	Other
Chalcedory	Travertine	
	Organic	Organic
Fossiliferous limestone		
Coquina	Fossil clam	
Radiolarian	Silica microfossils, chert	

Terrestrial sediments are material from the sediment load in rivers, dust storms, and turbidity currents. The physiography of the earth's surface has been tectonically shaped and modified by sedimentary, erosional, and depositional processes. These processes cause weathering and disintegration of rock masses at high topographic levels, and the removal of rock materials to lower levels where they are deposited, infilling depressions within them. The ocean basins form the ultimate base of the erosion. The present discussion addresses only modern marine environments where the sediments are normally unconsolidated. The generalized sediment distribution in the marine environment is shown in Table 5.2.

This table summarizes composition of deep-ocean sediments and the relative abundance of various types of deposits. Calcareous sediments are common as shown above at relatively shallow depths, and dissolve at approximately 4000 m. At great depths the predominant sediments type is brown clay (i.e., terrestrial clay).

The types of seascapes that develop and their sediments depend on the interplay of numerous factors of which the most important elements follow. The hydraulic regime includes weather-induced waves, currents and storms, tides, fronts between water masses, internal waves and tides, and upwelling. This movement of water masses determines the amount and rate of sediment erosion, transport and deposition, and sediment sorting. Sediment is supplied to the marine environment from a variety of sources: the continents by rivers, wind, and glaciers; offshore sources by waves, currents, tides, and gravity currents; or is biologically secreted or chemically precipitated from dissolved materials in the seawater. Fluctuations in sediment supply can result in major changes of the basin sedimentation pattern. The climate also partly controls the hydraulic regime in the sea and is the prime control on biological and chemical processes, especially in shallow water and emergent water areas

TABLE 5.2
Marine Sediment Distribution by Generalized Types

Continental shelf:
- Inner part: muds
- Tropic/subtropic: carbonates
- Temperate/polar: quartz and feldspar
- Outer part: sands

Continental slope: calcareous oozes

Ocean basins:
- Water depth <4–5 km: carbonate oozes
- Water depth >4–5 km: pelagic clays

Southern ocean and (Pacific) equatorial: siliceous oozes

A summary of the area and frequency of the ocean bottom occupied by these various sediments is presented in Figure 5.1.

Source: Richards (1981).
Notes:
(1) Siliceous material – radiolarian and diatomaceous.
(2) Calcareous – foraminiferous.

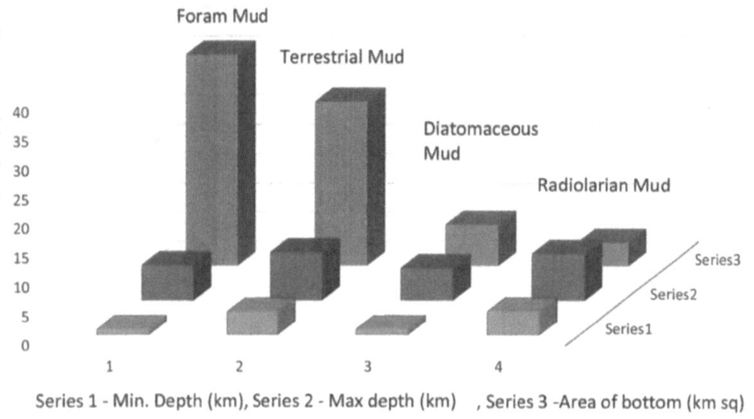

FIGURE 5.1 Distribution of deep ocean sediments with water depth (Adapted from Gross, 1967).

(e.g., lagoons and tidal flats). Sediment supply from the coastal plains is controlled by a combination of water, wind, and the amount of vegetation. The tectonic setting determines the type of the seabed bottom and its size, shape, and bathymetry, which in turn controls the hydraulic pattern. The tectonic activity of the basin influences subsidence rates, sea-level changes, and sediment supply. These factors combine to determine its long-term behavior. Sea-level fluctuation particularly affect the long-term behavior of the shoreline, which migrates landward during trangressions and seaward during regressions.

The sum of the physical, chemical, biological, and geomorphic conditions that operate in geographically restricted areas, collectively termed *sedimentary environment* determines the nature, or facies, of the sediments in these areas and their quantity and aerial distribution (Figure 5.1). The facies of a sediment is the sum of the lithological (compositional and mineralogical), physical, and biological aspects imparted to a sediment during and after its deposition.

5.2 CHARACTERIZATION OF SEDIMENTS

5.2.1 Introduction

The high variability of soils is reflected by the wide range of their geophysical, geochemical, and mechanical properties. Soils are conveniently sorted, or classified, into groups showing similar behavior or based on their significant properties. Soil classification has proved to be a valuable tool for the characterization and correlation of soils, and the efficient use and interpretation of both geological and engineering information. In general, soils (i.e., sediments) can be broken into two primary categories depending on whether they were transported into or were formed within the environment: (1) allochthonous and (2) antochthonous (Figure 5.2).

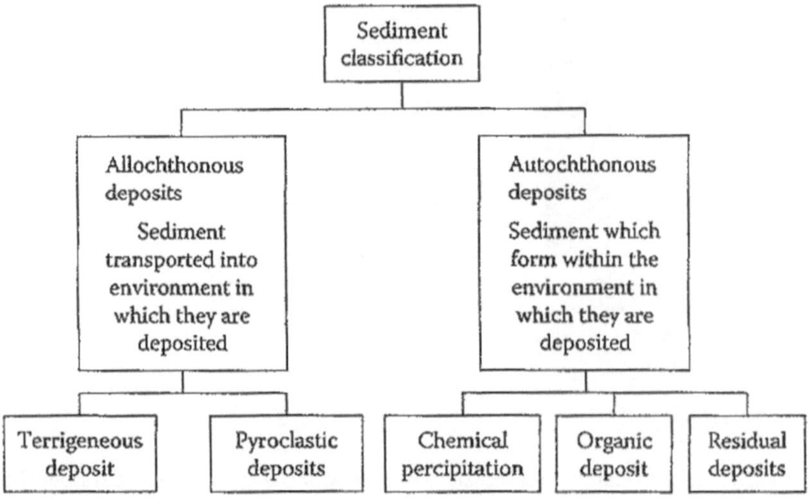

FIGURE 5.2 Basic division of sediments (Chaney and Almagor 2016).

5.2.2 Clay-Like Material

Physiochemical interactions are essentially electrical in character. Enumerated in a roughly decreasing order of strength they include repulsive forces (electrical repulsion between like-charged particles), covalent bonds (attraction between atoms resulting from the sharing of pairs of electrons), electrostatic interactions (attraction of unlike charges and repulsion of like charges as in ionic bonds), hydrogen bonds (attraction of hydrogen ions by electron-rich ions), and van der Waals forces (attraction between atoms or molecules due to their electrical asymmetry). Clay minerals are chemically stable due to the strong, partially covalent bonding between the tetrahedral silicon units and the octahedral aluminum- and magnesium-hydroxyl units within the clay sheets (Table 5.3).

TABLE 5.3
Properties of Major Types of Silicate Mineral

	Types of Clay		
Property	Montmorillonite	Illite	Kaolinite
Size (m)	0.01–1.0	0.1–2.0	0.1–3.0
Shape	Irregular flakes	Irregular flakes	Hexagonal crystal
Specific surface (m^2/kg)	700–800	100–120	5–20
External surface	High	Medium	Low
Cohesion, plasticity	High	Medium	Low
Cation exchange capacity (meq/100 g)	80–100	15–40	5–15

Source: Chaney and Almagor (2016).

These sheets are subjected, however, to intensive surface physiochemical interactions. Every clay particle carries an electrical charge, usually negative, arising from a number of sources. These sources are the following: the inexact charge balance on its outer faces (i.e., termination of the crystal lattice); isomorphous substitution and the random absence of cations in the crystal lattice; and, to a lesser degree, from ionization (dissociation) of surface groups (mostly hydroxyls), adsorption of anions from the surrounding solution, and the presence of organic matter. The smaller the particle the larger its specific surface and, consequently, its surface chemical activity as reflected by its cation exchange capacity, plasticity, and cohesion (Figure 5.3).

The specific surface is defined as the surface area of a particle divided by either its mass or volume. The behavior of clay particles, in the range from approximately 1 mm to 1 nm in size termed colloids, is mostly determined by surface-induced electrical forces (physiochemical interactions: flocculation and dispersion) rather than by mass-derived forces (gravity: settling). In contrast to clays, silts and sands with specific surfaces at least 10,000 times larger are chemically inactive. Clay particles carry a net negative charge on their surfaces. This is a result of both isomorphous substitution and a break in the continuity of the structure at its edges.

In order to achieve electro-neutrality, the clay particles attract positively charged ions or cations and polar water molecules from the surrounding solution onto their surfaces. This process is called hydration. Hydration occurs either directly onto the particle surface or indirectly, around the adsorbed cations, resulting in their phenomenal volume growth (sevenfold or more). Due to their size and the repulsion between the surface-attracted cations and water molecules, the hydrated ions move away from the particle surface to equilibrium positions where they best satisfy both surface attraction and cation repulsion forces between individual particles and their concentration decreases exponentially with distance from the particle surface (Figure 5.4). The curve shapes depend on the mineralogical nature and chemical composition of the particles.

The surface-held water is denser than the water of the surrounding solution, forming double layers (approximately 400 Å thick) around the clay particles. The water nearest to the surface of the clay particle surface (approximately 5 Å) is strongly attracted to it, gradually becoming free with distance away from the particle surface.

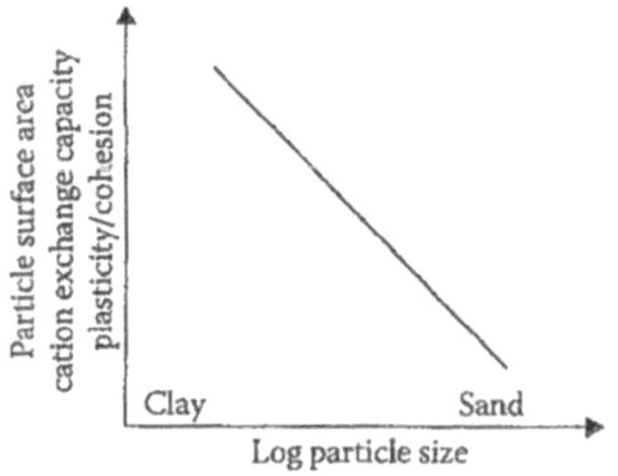

FIGURE 5.3 Particle surface area and other properties versus particle size.

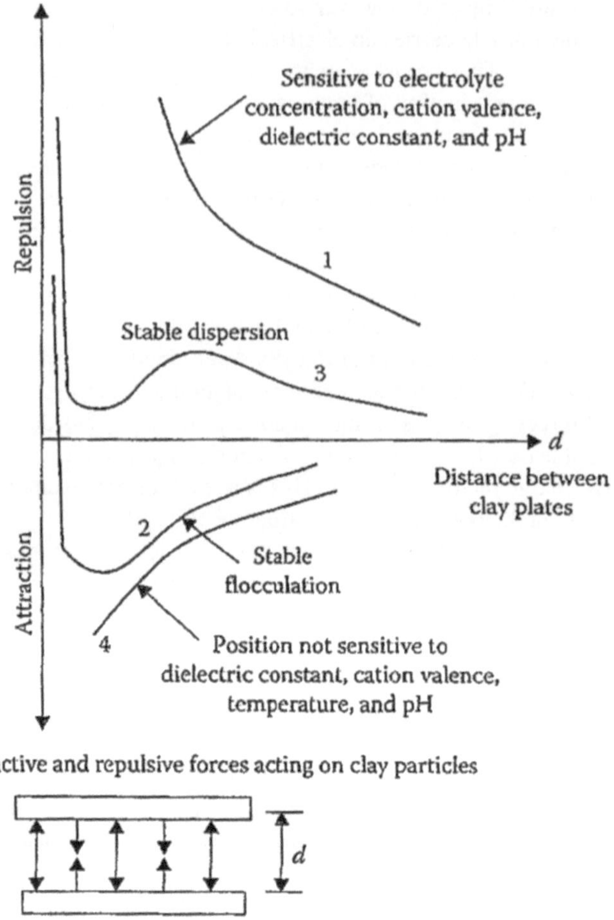

FIGURE 5.4 Relationship between the forces of particle interaction, F, and distance between particles.

At the margins of the double layers, the concentration of cations and water molecules becomes equal to that of the surrounding solution.

The cations are generally weakly held to the clay mineral surfaces, and can be readily displaced by other cations present in the solution. The cation (base) exchange depends on the ionic strength (charge to ionic radius ratio) and the concentration of the solution. The cation exchange (or adsorption) capacity of a clay mineral reflects its charge deficiency per unit mass and is a function of the particle composition and of its specific surface (Table 5.2).

Clay particles are held in suspension by electrostatic forces that act between their surface charge and that of the ions of the solution, thus resisting settling by gravity. The interaction between any pair of clay particles is governed by the net result of the balance between there, attractive forces primarily van der Waals, curve 4 (inversely proportional to the square of the separation distance), and the repulsive electrostatic forces curve 1 (decreasing exponentially with increasing distance of separation) (Figure 5.4). If the energy provided to the system (by the salt ions) is less than the maximum of the repulsive curve 3 in Figure 5.4 (the barrier energy), the particles will separate, or disperse, spontaneously with loss of the potential energy of the system. If, however, the energy required to bring the particles closer than the distance corresponding to the energy barrier is available, the particles will move rapidly toward each other, or flocculate. This movement will result in a release of energy, as the particles move to a separation distance dictated by the born repulsive forces between the clay particles, which correspond to the minimum of curve 2. Born repulsive force develops at contact points between particles, resulting from the overlap between electron clouds. It is sufficiently great to prevent the interpenetration of matter. At separation distances beyond the region of direct physical interference between adsorbed ions and between hydration water molecules, double-layer interactions provide the major source of interparticle repulsion. Particle separation is more difficult to achieve than particle aggregation, because the amount of energy required to separate the particles is much larger than the energy needed to bring them together. Dispersion in seawater occurs where water turbulence prevents contact between clay particles or in the presence of peptizers, such as dissolved phosphates carried by rivers. The tendency toward flocculation increases as the suspended particle concentration, the electrolyte concentration in the solution, the valence of ions, the temperature increase, and the dielectric constant (size of hydrated ions, pH, and anion adsorption) decrease. Likewise, water turbulence can increase interparticle collisions, thus providing the force required to overcome the potential energy barrier that prevents flocculation (Kranck, 1980). Flocculation is also enhanced in the presence of organic matter (>5%). Readily attached to silicate minerals and adsorbed by water molecules, organic matter effectively reduces the net surface charge of the clay particles. Thus, the presence of organic matter allows the clay particles to approach one another more closely, and in addition promotes particle-binding and bonding by polymeric products of microbial metabolism and within interwoven plant fibers (O'Brien, 1970; Pusch, 1973a, 1973b; van der Ven, 1981; Bennett and Hulbert, 1986). As more and more particles flocculate, the agglomerates become large and settle by gravity, forming a sediment on the seafloor in which the inter-particle pores are filled with seawater.

5.2.3 Clay Microstructure

The chemical nature of the fluid medium of saline water strongly affects the electrochemical interactions between suspended clay particles, thus determining the fabric characteristics of dispersion or flocculation prior to deposition. In non-saltwater the clay particles remain dispersed, settling slowly in quiet water and forming open networks of more or less uniformly distributed, largely parallel-oriented particles with moderately high void ratios. However, on contact with saline seawater the clay particles in suspension rapidly flocculate, producing voluminous agglomerates, several micrometers in width, of randomly oriented particles (flocs or floccules) (Figure 5.5). These flocs in turn tend to become denser and smaller as the water salinity decreases (Mitchell, 1956; Keller, 1957; Whitehouse et al., 1960; van Olphen, 1977). Particle associations in clay suspensions can be described as follows and as illustrated in

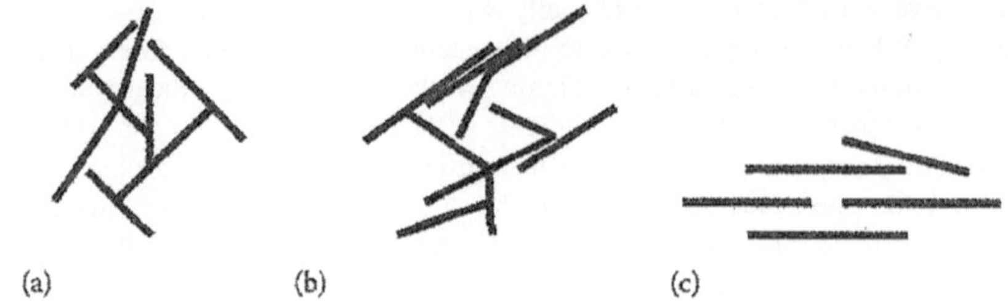

FIGURE 5.5 Various fine-grained sediment structures. (a) Undisturbed saltwater sediment (flocculated), (b) undisturbed freshwater sediment (partially flocculated), and (c) remolded sediment (dispersed).

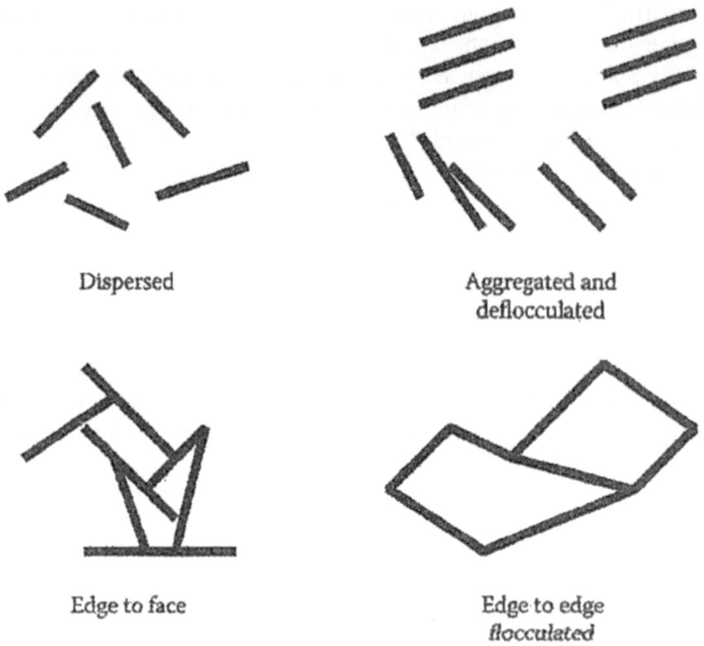

FIGURE 5.6 Modes of particle association in clay suspension and terminology.

Figure 5.6 (van Olphen, 1977): 1. Dispersed – No face-to-face (FF) association of clay particles, 2. Aggregated – FF association of several clay particles.

5.2.4 Texture, Structure, and Composition of Soil Particles

The properties of a soil are a function of both its texture and structure, and the chemical and mineralogical composition of its particles. Some of these properties are its capacity to adsorb water, its plasticity and cohesion, and its changes in mechanical behavior due to variations in environmental conditions and time, such as its compressibility, permeability, strength, and stress transmission. The sediments together with fossils can also provide data needed for the reconstruction of the geological processes that have formed the soil. Texture is concerned with the size of the soil's minerals and rock particles, specifically referring to the relative proportions of particles of various sizes. Structure refers to the distribution and orientation of the particles within the soil mass, and their arrangement into groups or aggregates. These arrangements are seen in hand specimens and soil sections, or by ultra-thin methods. Large features of sedimentary origin, such as stratification, bedding, lamination, ripple and wave marks, unconformities, and deformations of various origins, define sedimentary structures. Composition refers to the nature and arrangement of the atoms in a soil particle. The texture, structure, and composition, and consequently the physical properties, of detrital terrigenous and volcanic (lithogenous) sediments, and of sea-born biogenic (biogenous) and chemical (hydrogenous) sediments are very different. The particles of terrigenous and volcanic sediments produced by physical and chemical weathering processes are silicate and aluminosilicate minerals. These particles (except glacial rock floor and those recently introduced by submarine volcanic eruptions) are very resistant to both physical and chemical weathering. The size distribution of these sediments is not subject to ready change, and they remain unconsolidated and chemically unaltered in their depositional environments, even if they become subaerially exposed during sea level falls. By contrast, the biogenous and hydrogenous sediments, which are mostly from shallow water environments, are subject to intensive diagenesis. These sediments commonly become cemented on the seafloor, and during exposure they readily alter chemically and mineralogically into typically either carbonates or evaporites. Such changes greatly affect the texture, structure, and composition, and consequently the physical properties of sediments. The relative abundance of clay minerals in the northern South China Sea and adjacent rivers is presented in Figure 5.7.

5.2.5 Clastic Soils

5.2.5.1 Size, Shape, and Structure Disintegration

Size, shape, and structure disintegration caused by impact, abrasion, grinding, and transportation of the sediment solution to the sea produce discrete soil particles widely ranging in size, shape, and texture. Soil particles vary in size from 1 nm (1 × 10E−8 m = 10 Å) up to large rocks several meters in diameter, a range of one to more than a billion. Soil particles are conveniently separated into groups according to size, their ranges arbitrarily set, and each assigned a name. There are no universally accepted definitions of size fractions, and different size classifications. Sediment transported into an environment in which they are deposited are developed for different uses (e.g., geological, pedological, and agricultural soil engineering). Figure 5.8 presents the size definitions widely used in geotechnical work. The degree of similarity in size of the particles in sediment is termed sorting. Engineers consider a soil poorly sorted if it is predominately of one particle size. By contrast, geologists consider a soil well sorted if it is predominately of one particle size. In this book we will use the engineering interpretation. The particle sizes, determined by mechanical size analyses, are plotted as a summation or frequency curve, the form of which gives more information that can be readily visualized than from a table of percentages, for example, the degree of sorting or grading.

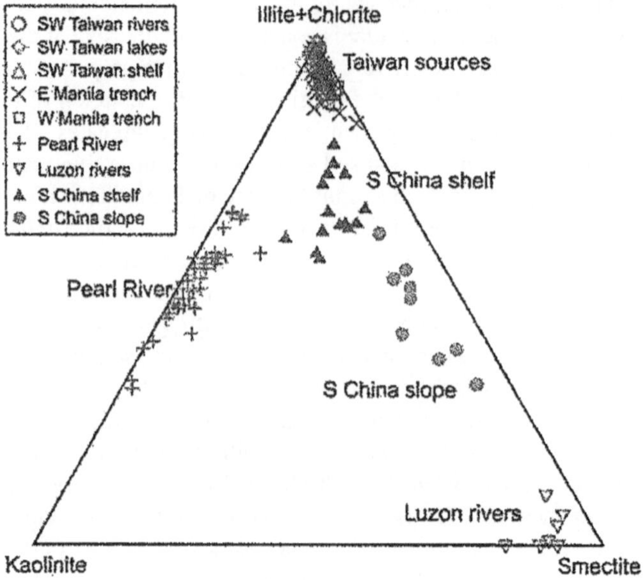

FIGURE 5.7 Comparison of clay mineral assemblages among south-western Taiwan, the Pearl River, Luzon, and the northern SCS (Liu et al., 2008). Reprinted with permission of Elsevier.

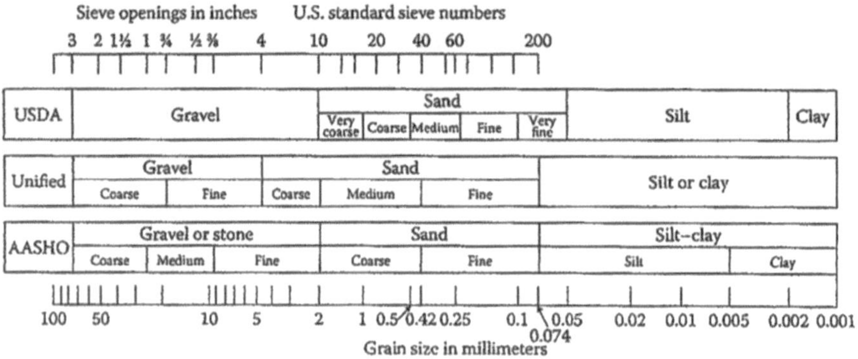

FIGURE 5.8 Comparison of particle size scales. (Data from Bureau of Reclamation, Earth Manual, Part 1. 3rd Edition, US Department of the Interior, US Government Printing Office, Washington, DC, 329, 1998.)

The very coarse particles, from gravel upward, are fragmented rocks and are usually highly variable in shape. They may be irregularly shaped, more or less rounded, or even flat. Fragmented rocks are usually a minor constituent and rated separately in soil because of their size.

The sand-, silt-, and clay-sized components of a soil may be of diverse origins and compositions. Although the shape of the individual particles may vary in a given soil, they are basically the same in each of these major size groups. Usually, the shape and size of the particles refers to the composition and the crystal structure of the grains. Sands and silts tend to be bulky, and are approximately equidimensional, reflecting the physical weathering process experienced during their transportation. Clays, the result of chemical weathering, typically possess platey, sheet like shapes reflecting their crystal structures. The sphericity, or the degree of closeness to a sphere, is a measure of the grain shape. The degree of roundness (or that of angularity) refers to the sharpness of the edges and corners of a particle in the silt and larger size range. A high degree of angularity of particles is indicative of recent fragmentation of a larger rock fragment, short transportation, or high resistivity to weathering, whereas a high degree of roundness reflects the opposite. Roundness and sorting are normally closely related. Well-sorted deposits of coarse; well-rounded grains are typical products of winnowing and abrasion encountered in high-energy environments, such as beaches; while deposits in low-energy environments are likely to consist of more poorly sorted material, with a fine grain size and more angular particles. The degree to which a deposit is both well-sorted (i.e., contains a variety of different particle sizes) and rounded defines its textural maturity (Folk, 1951).

5.3 CLASSIFICATION OF SEDIMENTS

5.3.1 Systems of Classification

Soil classification deals with the categorization of soils based on distinguishing characteristics as well as potential criteria that dictates choices in use. Soil classifications arise from the need for (1) rational, yet convenient characterization of soils; (2) their correlation differentiation "comparison" and contrast; and (3) methods for effective communication based on a uniformity of criteria and nomenclature.

It is difficult to classify soils because their gradations differ from one form to another. This difference is caused by a combination of the frequent mixing of sediments of different *size*, composition, and origin, and changes caused by a variety of transportational and diagenetic processes. Therefore the existing classifications are constructed on the basis of a few significant properties of the sediments. Several classification systems of sediments have been proposed based on their chemical and mechanical composition, selected geotechnical properties, agencies of transport and deposition, sedimentary environments, and so on. Difficulties, however, still remain, mostly because the significant properties of one sediment group according to any classification system are not necessarily the significant properties of another. Even very partial classifications based on a single, well-defined significant property (e.g., permeability, compressibility) evaluated to address specific types of problems have serious limitations. For example, the most commonly used classification system based on grain size does not consider the chemical and mineralogical composition of the soil particles even though these properties govern particle hardness and density, reaction to pore fluids, and so on. Consequently, unified classification systems evaluated to simultaneously employ a wide range of significant properties must either employ descriptive terms (e.g., the General Classification of Marine Sediments – Figure 5.9), or be accompanied by auxiliary descriptive charts (as is done in the Unified Soil Classification System [USCS], Wagner [1957]).

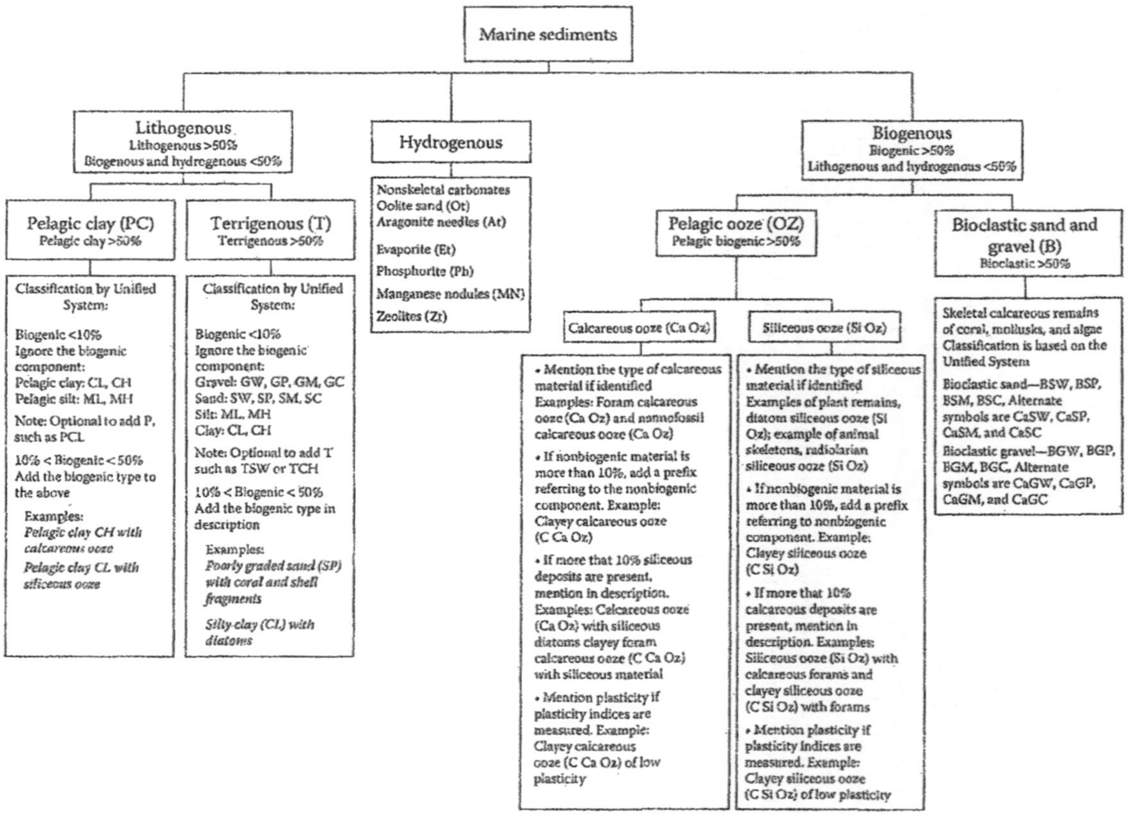

FIGURE 5.9 Chart for classification of marine sediments (From Noorany, 1983).

In a geological classification system, the composition, texture, and structure of a sediment and the fossils it contains are related to origin, aiming at the reconstruction of the geological history: generic formation (extrabasinal-terrigenous, clastic sediments, and intra-basinal-in-place formed biogenous and chemical or hydrogenous sediments), agents of deposition (transportation of sediments by the action of waves, currents, tides, gravity, ice, winds, and organisms), and sedimentary environments (e.g., nearshore, shallow sea, continental slope, pelagic, and glaciomarine environments). In the first classification system, the clastic sediments are subdivided into textural classes (i.e., gravel, sand, silt, clay), and the biogenic and chemical sediments are classified according to their major chemical or mineralogical constituents.

A principal objective of engineering classifications is to allow a reasonable assessment of the fundamental properties of a given soil by comparison with another that has similar behavior. This comparison allows the reduction or the better direction of complex testing procedures that are employed to evaluate them. Therefore a classification system must employ simple, easily obtained criteria, such as those evaluated by simple index tests. It will lose its value if the tests become more complicated than the tests to directly measure the fundamental properties. Casagrande (1948) provided some guidelines for classification systems, which are as follows: (1) properties in undisturbed condition, (2) indication of properties, (3) applicability in both visual and laboratory references, and (4) simple system of notation. The methods of classifying soils for engineering purposes are based on either the properties of raw materials of soils, or on properties of the soil in its undisturbed condition. Within these two basic methods, Casagrande (1948) outlined the various techniques available: (1) properties of raw material of soil, and (2) Properties of soil in its undisturbed state. Under properties of raw materials of soils, there are (1) textural soil classifications and (2) road subgrade classifications (such as American Association of State Highway and Transportation Officials and airfield subgrade classifications).

5.3.1.1 Engineering Classification of Siliciclastic Sediments

Soil characteristics most used in engineering classification systems are particle size and consistency (Atterberg limits). These simple tests reflect the soil's physical properties, composition, moisture content, and so on. The information obtained by these tests enables an individual to assess the quality of a soil for construction or as a foundation. Several *systems* are in use in the United States and throughout the world, which have application in the marine environment. These are the USCS, which was originally proposed by A. Casagrande in 1942 and was later revised and adopted in 1952 by the US Bureau of Reclamation and US Army Corps of Engineers and later by the American Society for Testing and Materials System (ASTM) D2487. In one form or another the unified system is used for virtually all geotechnical work throughout the world. In these systems soils are classified by the dominant particle size and organic content into gravels (G), sands (S), silts (M), clays (C), organic sediments (0), and peats (Pt). Coarse-grained or granular soils (G and S) are further sub-divided by grading, and fine-grained or cohesive soils (M, C, and 0) by liquid limit and plasticity index, into 15 groups, which have distinct engineering properties.

One of the most useful methodologies of classifying sediments is by using the end member concept (i.e., sand, silt, clay) (Shepard, 1954). Sediments containing three

TABLE 5.4
Deep Sea Sediment Classification System

Terrigenous sediment	Less than 30% carbon carbonate and silica
Biogenic sediment Calcareous Sileous	Greater than 30% biogenic calcium carbonate and silica
Chemogenic sediment	Chemical precipitates from seawater
Volcanogenic	Composed mostly of pyroclastic material
Polygenic	Red clay

Source: Adapted from Lisitzin (1972).

constituents can be classified in a triangular diagram in which each apex represents 100% of one of the three constituents (i.e., sand, silt, and clay). More general classification schemes based primarily on sediment composition but also on origin have *also* been presented. Lisitzin (1972) proposed a system in which he divided sediments into terrigenous, biogenic, chemogenic, volcanogenic, and polygenic (red clay) (refer to Table 5.4).

Berger and von Rad (1972) followed this effort with a classification system for the predominate deep-sea sediments: pelagic and hemipelagic materials (Table 5.5). Hemipelagic sediments are not deposited as slowly as pelagic sediments. They are deposited on continental shelves and rises, and ordinarily accumulate too rapidly to react chemically with seawater. Berger's classification system does not extend to a variety of other sediments. The DSDP (deep-sea drilling project) has also developed a couple of classification systems. During the first three phases of the DSDP from 1968 to 1976 the description and classification evolved from a poorly defined qualitative system to a more rigorous system.

This new system provided sediment names and classes based on standardized descriptive parameters (van Andel, 1981). A Figure 5.11 summary of the procedures used in the application of this classification have been presented by a number of authors (Hayes et al., 1975; Weser, 1973) (Table 5.5). This approach was revamped in 1974 to simplify sediment names and promote more uniform usage to decrease errors, as shown in Figure 5.10.

This methodology can also be used to classify a soil comprised of mixed sediments, non-clastic sediments, environment-dominated sediments (Figure 5.11), or other sediment types, provided they possess a common parameter that can be expressed quantitatively (e.g., percentage). Four-component sediment systems can be plotted in three dimensions within the faces of a tetrahedron.

5.3.1.2 Classification of Carbonate Sediments

In the following discussion, only calcareous sediments, which form the bulk of the modern marine carbonate sediments, will be addressed. Dolomites (Ca-Mg carbonates), normally the products of long diagenesis of limestones, are usually excluded from engineering classifications of carbonate sediments, because only minor quantities

TABLE 5.5
Classification of Deep-Sea Sediments

I. Pelagic deposits (oozes and clays)
   <25% of fraction <5 μm is of terrigenic, volcanogenic, or neritic origin
   Median grain size <5 μm (except authigenic sediments and pelagic fossils):
   A. Pelagic clays. $CaCO_3$ and siliceous fossils <30%
      1. $CaCO_3$: 1%–10% (slightly) calcareous clay
      2. $CaCO_3$: 1%–30%; very calcareous (or marl) clay
      3. Siliceous fossils: 1%–10% (slightly) siliceous clay
      4. Siliceous fossils: 10%–30%; very siliceous clay
   B. Oozes, $CaCO_3$ or siliceous fossils >30%
      1. $CaCO_3$; >30%; <2/3 $CaCO_3$; marl oozes; >2/3 $CaCO_3$; chalk ooze
      2. $CaCO_3$; <30%; >30% siliceous fossils: diatom or radiolarian ooze
II. Hemipelagic deposits (muds)
   >25% of fraction: >5 μm is of terrigenic, volcanogenic, or neritic origin
   Median grain size > 5 μm (except authigenic minerals and pelagic fossils):
   A. Calcareous muds. $CaCO_3$ > 30%
      1. <2/3 $CaCO_3$; marl mud; >2/3 $CaCO_3$; chalk mud
      2. Skeletal $CaCO_3$ >30%: foram, nanno; coquina
   B. Terrigenous mud. $CaCO_3$ <30%. Quartz, feldspar, mica dominant
      Prefixes; quartzose, arkosic, micaceous
   C. Volvanogenic muds. $CaCO_3$ <30%. Ash, palagonite, and so on dominant
III. Pelagic and hemipelagic deposits
   A. Dolomite-sapropelite cycles
   B. Black (carbonaceous) clay and mud: sapropelites
   C. Silicified claystones and mudstones: chert
   D. Limestone

Source: Berger and von Rad (1972).

of present-day marine dolomite are inorganically deposited and diagenetically formed in a few hot, shallow water, and Mg-rich lagoons (e.g., the Persian Gulf, South Australia). Carbonate (and chemical) sediments cannot be classified on the basis of their texture alone for the following reasons: (1) Carbonate matter is subjected to intensive diagenesis: dissolution, aggregation, agglutination and pelletization, chemical and mineralogical alteration, carbonate precipitation, encrustation, cementation, and recrystallization to highly variable degrees, especially during periods of exposure. Aragonite crystals (lime mud) infill interstices among coarse-grained particles and cavities in reefs. The crystals are protected in these locations from winnowing even in highly agitated water environments. The aragonitic crystals are products of disintegration of skeletal debris and are particularly prone to diagenesis processes; and (2) in most places carbonate sediments are variously mixed with siliciclastic sediment. Geologists have classified carbonates according to (1) origin (e.g., skeletal or non-skeletal, or even more precisely, foraminiferal, coralline, algal, diatomaceous, etc.),

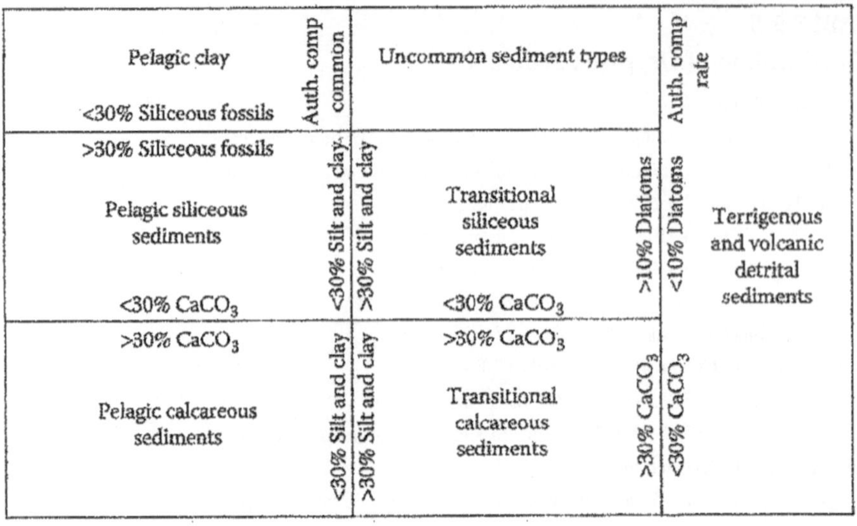

FIGURE 5.10 Summary diagram of JOIDES sediment classification employed after Leg 38 of the DSDP (After Van Andel, 1981).

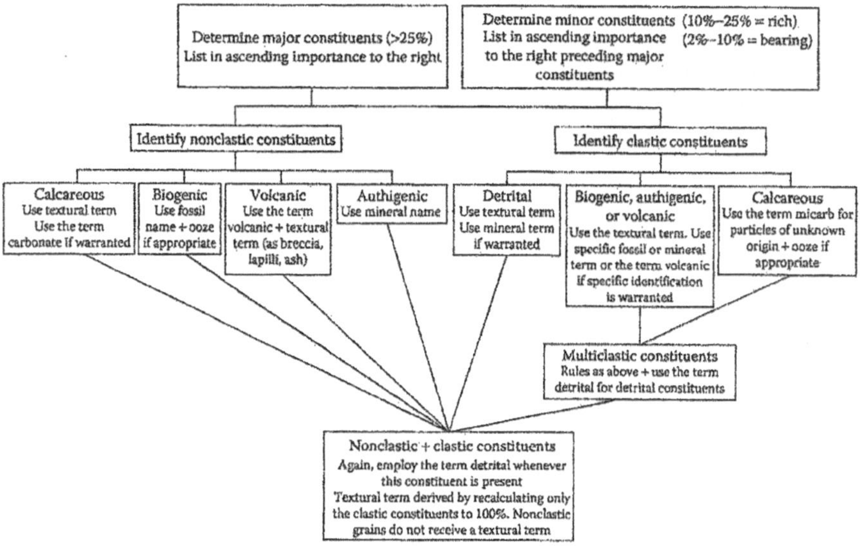

FIGURE 5.11 Summary diagram of JOIDES sediment classification employed after Leg 38 of the DSDP (After Van Ardel, 1981).

(2) mineral content (e.g., aragonitic, calcitic, etc.), and (3) depositional environments (e.g., pelagic, littoral, peritidal, etc.), among other methods, which aid in their interpretation of geological phenomena. All these classifications essentially employ particle-size terms: either directly by use of textural categories (i.e., gravel, sand, silt, and clay (Chaney et al., 1982; Figure 5.12) or numerical dimensions; or indirectly by use of environmental terms (such as pelagic, littoral, etc.) and biologic terms (such as foraminiferal, nannofossil, mega-skeletal, etc.), which are indicative of size. Textural classifications of carbonate sediments are very useful to geologists, because they refer to the hydraulic regime responsible for their accumulation in a particular depositional environment. The redistribution of carbonate particles by waves, currents, tides, and gravity is a function of their size, even though it primarily reflects the size of skeletons and precipitant grains. Various scales for the degree of indication of carbonate sediments were incorporated into most of the carbonate classifications, roughly expressing the amount of diagenesis (cementation and recrystallization) that they experienced.

The classifications of carbonate sediments currently in use are based on the relative quantities of the infilling materials with respect to the coarse-sized particles (sand and gravel) (Folk, 1956, 1959; Dunham, 1962; Leighton and Pendexter, 1962; Fookes and Higginbottom, 1975, Clark and Walker, 1977). The former are defined by their state and degree of cementation; discrete, tiny particles (micrite) that may have transformed into mosaic-like cement (sparite). The latter are identified by their origin (fragmental intraclasts, ooids, skeletal debris, and fecal pellets). The sediments thus classified represent a continuous sedimentary sequence from a sediment that is made purely of lime mud (micrite) to limestones consisting of a rigid organogenic framework (reefal rocks-coralline, bryozoan, algal, etc.). These classifications bear genetic implications because the fine-to-coarse particle ratio is controlled to a large degree by sorting processes and the availability (productivity) of the coarse particles and the lime muds in a given depositional environment. At present there is no engineering method that is specifically capable of classifying calcareous materials. A variety of parameters have been proposed to characterize the engineering properties of marine carbonate sediments (refer to Table 5.6). Some of these characteristics are listed as follows:

1. Carbonate content can range in value from 0% to 100% by dry weight and is quickly and easily measured by acid treatment methods.
2. Atterberg limits when used with a plasticity chart provide information on the non-carbonate fraction.
3. Dissolution (mostly in sediments that were exposed subaerially) is highly dependent on climate, and results in cavernous structures and relative concentration of solution-resistant, residual deposits (i.e., mostly clays and metallic oxides).
4. *Cementation* may vary from a strongly cemented limestone to a soft chalk that can be scratched with a thumbnail, to loosely aggregated particles of various sizes, to clastics that range from very large bioclastic debris to fine-grained oozes with no cementation (lime muds).
5. *Grain size* may reflect the size of a cemented mass or individual particles that comprise a matrix.
6. *Geological classifications* is based on mineralogy, origin, and biologic/biogenic composition.

Name	Shape	Smallest subparticle shape	Depth location (m)	Environment High-energy area	Environment Low-energy area	Porosity	Overall material behavior[a]	Chemical constituent
Foraminiferan shells (tests)	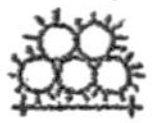 1 mm		3500	Benthic forams	Planktonic forams	Porous (porcellaneous)	Cohesionless (sandy silt)	Calcite (Mg-calite in small quantities)
Pteropod shells (tests)	 1–2 mm	No discrete subparticles	3000	Tropical and subtropical waters	Globigerina	Porous (porcellaneous)	Cohesionless (sand)	Aragonite
Coccolithic plants (nannofossils)	 0.01 mm		3500–5000	—	Mainly latitudinal variation of species type	No inherent porosity (hyaline)	Cohesionless (silt)	Calcite shells of the algal family; skeletons may be protected by organic or $MgCO_3$ coatings
Corals	 20 cm	No discrete subparticles	0–35	Branching, massive and encrusting corals	Foliose and encrusting corals	Porous (porcellaneous)	Cohesionless (sandy gravel)	Mg-calcite aragonite
Precipitate	 0.1–0.6 mm Highly variable	No discrete subparticles	35–200	Oolites	Pelletoids	No inherent porosity (hyaline)	Cohesionless (sand)	Mg-calcite
Benthic materials	—	—	35–200	Branching coraline algae, forams, planktonic feeders	Mollusks, benthic forams, echinoid fragments	—	—	Mg-calcite

[a] Qualitative description of material response (i.e., sand, sandy silt, sandy gravel, silt) based on assumed particle size.

FIGURE 5.12 Summary of marine carbonate sediments (From Chaney et al. 1982). Reprinted with permission of ASTM.

TABLE 5.6
System for Describing Calcium Carbonate Material

Description

1. Cementation
   a. No cementation
   b. Weak cementation
   c. Strong cementation
      (i) Unknown – The soil has a soft rock-like appearance. Unconfined compressive strength should be indicated.
      (ii) Partial – The soil contains cemented aggregates. This should be noted grain size distribution (GSD).
2. Grain size distribution and plasticity – For strongly cemented soils GSD is not very relevant. For relevant uniform cementation – size of constituent particles should be indicated, for partial cementation – GSD of soil after removing aggregates should be indicated and the size and proportion of aggregates noted separately.
   a. Plasticity – For fine-grained soils in which intraparticle voids cause error in GSD and Atterberg limits, field classification procedures may be used for providing the relevant information in a qualitative sense.
3. Nature of carbonate component
   a. Carbonate content – Soils having more than 30% carbonate content should be termed as carbonate soils.
   b. Particle size of carbonate material – The carbonate content in the sand and in the silt-clay fractions should be determined separately and indicated. Microscopic studies mentioned below will also give information about particle size.
   c. Particle characteristics and origin – Microscopic studies – optical microscope for sands and scanning electron microscope for fine grained soils should be conducted. Presence of thin-walled material and intraparticle voids should be highlighted.
   d. Mineralogy – X-ray diffraction analysis should be performed.
   e. Geologic name – If possible to identify, the geologic name may be indicated.
4. Nature of noncarbonated component – Information on noncarbonated material is determined by dissolving the carbonate material in HCL, separating the remaining soil and conducting the following tests on it.
   a. Particle size – Grain size distribution analysis
   b. Particle characteristics – Microscopic studies
   c. Mineralogy – X-ray diffraction analysis

Source: Datta et al. (1982).

In general, it appears that all the above methods taken together provide an excellent means of characterizing a calcareous material. In a similar fashion, engineers might use several existing classification systems to adequately characterize the properties, behavior, and expected performance of calcareous soils, since it is unlikely that any single system will perform all of these functions. Fookes and Higginbottom (1975) were perhaps the first to attempt classification of carbonate soil from an engineering viewpoint. They used four criteria for classifying carbonate sediments: (1) carbonate content, (2) degree of induration, (3) particle size, and (4) origin of carbonate material. Their classification system covers all types of carbonate sediments – from carbonate mud to carbonate gravel, from unindurated carbonate soil to hard carbonate rock, and from impure carbonate sediments to pure carbonate sediments. In this scheme, they separate unindurated from indurated sediments. The boundary between unindurated and induration is arbitrary. This arbitrary degree of induration is meant to be at the stage in induration where the porosity is nearly zero, due to cementation and re-crystallization, so that grain size is no longer a useful parameter in the mud-to-fine sand range. The transition between cementation and re-crystallization is inversely related to porosity. This can also be considered a transition between weak rocks and moderately strong rocks. Fine-grained carbonate sediments are termed lime mud instead of clay, because fine carbonate particles have very different properties than true clay minerals. The term *limestone* is reserved for carbonate rocks that are moderately strong.

Composition is used as a further subdivision of sands and gravels. Sands are divided by organic (bioclastic) versus inorganic (oolitic) content. When both are together, the proper term is oolitic bioclastic sand. The gravel range is divided by the degree of induration into gravels (unindurated), conglomerates or breccias (slight induration), and conglomerate limestone (indurated). Each of these divisions is subdivided based on the original materials. They are either organic shells, corals, or algae; or they are inorganic pisolites (large ooids over 2 mm in size). Clark and Walker (1977) slightly modified this scheme of classification using the same four variables and presented it in a more systematic manner.

REFERENCES

ASTM D2487, Standard Practice for Classification of Soils for Engineering Purposes (Unified Soil Classification System), 12pp.

Bennett, R.H. and Hulbert, M.H. (1986). *Clay Microstructure.* Geological Sciences series, International Human Resources Development Corporation, Boston, MA, p. 163.

Berger, W.H. and von Rad, U., 1972. "Cretaceous and Cenozoic sediments from the Atlantic Ocean," in D.E. Hayes, A.C. Pimm, et al. (Eds.), *Initial Reports of the Deep Sea Drilling Project*, Vol. 14, US Government Printing Office, Washington, DC, pp. 787–954.

Casagrande, A. (1948) "Classification and identification of soils," *Transactions of the American Society of Civil Engineers*, 113: 901–931.

Chaney, R.C. and Almagor, G. (2016). *Seafloor Processes and Marine Geotechnology*, Taylor & Francis, New York, 558pp.

Chaney, R.C., Slonim, S.M., and Slonim, S.S. (1982). "Determination of calcium carbonate contents in soils," In K.R. Demars, and R.C. Chaney (Eds.), *Geotechnical Properties, Behavior and Performance of Calcareous Soils*, Special Technical Publication 777, American Society for Testing and Materials, Philadelphia, pp. 3–15.

Clark, J.I. and Walker, B.F. (1977). "A proposed scheme for classification and nomenclature for use in engineering descriptions of Middle Eastern sedimentary rocks," *Geotechnique*, 27: 93–99.

Datta, M. et al. (1982). "Engineering behavior of carbonate soils of India and some observations on classification of such soils," in K.R. Demars and R.C. Chaney (Eds.), *Geotechnical Properties, Behavior, and Performance of Calcareous Soils*, ASTM STP 777, American Society for Testing and Materials, Philadelphia, PA, pp. 113–140.

Dunham, R.J. (1962). "Classification of carbonate rocks according to depositional texture," in W.E. Horn (Ed.), *Classification of Carbonate Rocks*, Memoir 1, American Association of Petroleum Geologists, Tulsa, OK, pp. 108–121.
Folk, R.L. (1951). "Stages of textural maturity in sedimentary rocks," *Journal of Sedimentary Petrology*, 21: 127–130.
Folk, R.L. (1956). "The role of texture and composition in sandstone classifications," *Journal of Sedimentary Petrology*, 26: 166–171.
Folk, R.L. (1959). "Practical petrographic classification of limestones," *American Association of Petroleum Geologists Bulletin*, 43: 166–171.
Fookes, P.G. and Higginbottom, I.E. (1975). "The classification and description of nearshore carbonate sediments for engineering purposes," *Geotechnique*, 25: 406–411.
Gross, M.G. (1967). *Oceanography: A View of the Earth*, United States Department of Energy, N.p., web.
Hayes, D.E., Frakes, L.A. et al.(1975). *Initial Reports of the Deep Sea Drilling Project*, Vol. 28, U.S. Government Printing Office, Washington, DC.
Keller, W.D. (1957). *The Principles of Chemical Weathering*, Lucas Brothers, Columbia, MO, p. 111.
Kranck, K. (1980). Experiments on significance of flocculation on the settling of fine grained sediments in still water. *Canadian Journal of Earth Sciences*, 17: 1517–1526.
Leighton, W.M. and Pendexter, C. (1962). "Carbonate rock types," in W.E. Horn (Ed.), *Classification of Carbonate Rocks*, Memoir 1, American Association of Petroleum Geologists, Tulsa, OK, pp. 31–61.
Lisitzin, A.P. (1972). *Sedimentation in the World Ocean*, Special Publication 17, Society of Economic Paleotologists and Mineralologists, Tulsa, OK.
Liu, Z.F., Colin, C., Li, X.J., Zhao, Y.L., Tuo, S.T., Chen Z. et al. (2010) "Clay mineral distribution in surface sediments of the northeastern SCS and surrounding fluvial drainage basins," *Marine Geology*, 277: 48–60.
Liu, Z. et al. (2008). "Detrital fine-grained sediment contribution from Taiwan to the northern South China Sea and its relation to regional ocean circulation," *Marine Geology*, 255: 149–155.
Mitchell, J.K. (1956). "The fabric of natural clays and its relation to engineering properties," *Highway Research Board Proceedings*, 35: 693–713.
Noorany, I. (1983). *Classification of Marine Sediments, Department of Civil Engineering*, Soil Mechanics Series, San Diego State University, San Diego, CA, 27pp.
O'Brien, N.R. (1970). "The fabric of shale – An electron microscope study," *Sedimentology*, 15: 229–246.
Pusch, R. (1973a). *Influence of Organic Matter on the Geotechnical Properties of Clays*, National Swedish Building Research Summaries, Stockholm, Sweden, p. 64.
Pusch, R. (1973b). Influence of salinity and organic matter on the formation of clay microstructure. In *Proceedings of the International Symposium of Soil Structure*, Gothenburg, Sweden, Swedish Geotechnical Society, Stockholm, Sweden, pp. 165–175.
Richards, A.F. (1981). Personal communication.
Shepard, F.P. (1954). "Nomenclature based on sand-silt-clay ratios," *Journal of Sedimentary Petrology*, 24: 151–158.
Sowers, G.B. and Sowers, G.F. (1961). *Introductory Soil Mechanics and Foundations*, 2nd edn. Macmillan, New York, p. 386.
Van Andel, T.H. (1981). "Sediment nomenclature and sediment classification during phases I-III of the deep sea drilling project," In G.R. Heath (Ed.), *Deep Sea Drilling Project Initial Reports Methodology*, U.S. Government Printing Office, Washington, DC.
Van der Ven, T.G.M. (1981). "Effects of polymer bridging on the selective shear flocculation," *Journal of Colloid and Interface Science* 81: 290–291.

Van Olphen, H. (1977). *An Introduction to Clay Colloid Chemistry*, 2nd edn. John Wiley & Sons, New York, p. 318.

Wagner, A.A. (1957). The use of the Unified Soil Classification system by the bureau of reclamation. In *Proceedings of the 4th International Conference on Soil Mechanics and Foundation Engineering*, London, Vol. 1, p. 125.

Weser, O.E. (1973). "Sediment classification," in L.D. Kulm, and R. von Hueme et al. (Eds.), *Initial Reports of the Deep Sea Drilling Project*, U.S. Government Printing Office, Washington, DC, Vol. 18, pp. 9–13.

Whitehouse, U.G., Jeffrey, I.M., and Debbrecht, J.D. (1960). Differential settling tendencies of clay minerals in saline water. In *Proceedings of the 7th National Conference on Clays and Clay Minerals*, Washington, DC.

6 Index, Compressibility, and Strength of Marine Sediments

6.1 INTRODUCTION

The physical properties of marine sediments vary from those of terrestrial origin to those that are unique to the marine environment. Typical properties of surficial seafloor soils are described by Demars and Anderson (1971), Sverdrup et al. (1942) and Herrmann et al. (1972). Table 6.1 summarizes properties of typical seafloor soils, as compiled by Herrmann, et al. (1972). A review of Table 6.1 shows that shear strength ranges approximately from 0.04 to 2.5 psi, while water content varies from 50% to 300%. Unit weights range approximately from 78 to 109 pcf. The test methods to obtain the various soil parameters presented in this chapter are discussed in detail in texts such as Chaney and Almagor (2016) and will not be presented here. In this chapter the various characteristics of these sediments will be presented.

Terms commonly used in defining characteristics of sediments are presented in the following abbreviated list of simplified definitions for the readers (ASTM D653). This list is followed by a discussion and typical data for three different soils: (1) clays, (2) calcareous, and (3) organic.

Consistency—The relative ease with which a soil can be deformed.

Compressibility—Property of a soil or rock pertaining to its susceptibility to decrease in volume when subjected to load.

Consolidation—The gradual reduction in volume of a soil mass resulting from an increase in compressive stress.

Density—The mass per unit volume

Dry density—The mass of dry soil or rock per unit total volume.

Dry unit weight—The dry density multiplied by standard acceleration of gravity

Liquid limit (w_L)—The water content, in percent, of a soil at the arbitrarily defined boundary representing the transition from the semi-liquid to plastic states.

Liquidity index (I_l)—the ratio of: (1) the water content of soil at a given condition/state minus its plastic limit, to (2) its plasticity index. This index presents the relative consistency of a cohesive of a cohesive soil in the natural state. For sensitive clay $I_L > 1$, soil deposits that are heavily over consolidated $I_L < 1$.

Permeability—The rate of discharge of water under laminar flow conditions through a unit cross-sectional area of a porous medium under a unit hydraulic gradient and standard temperature conditions (usually 20°C).

Plastic limit (w_p)—The water content, in percent, of a soil at the boundary representing the transition from the plastic to semi-solid states

TABLE 6.1
Engineering Properties of Various Sediment Types

Sediment Type	Shear Strength (psi)	Water Content (%)	Wet Unit Weight (pcf)
Terrigenic	0.04–2.5[a]	50–100	94–109
Red clay	<0.5	Extremely variable As high as 300	94–109[b] 78–94[c]
Siliceous	0.5–1.5	50–100	<94
Calcareous	0.5–1.0	100–200	94–109[b] 78–94[c]

Source: After Herrmann et al. (1972). US Government.
Notes:
[a] Among terrigenic sediments there are granular materials that have high strengths which cannot be stated in terms of shear strength
[b] Atlantic sites
[c] Pacific sites

Plasticity index (PI)—The range of water content over which a soil behaves plastically. Numerically, it is the difference between the liquid limit and the plastic limit.

Shear strength—The maximum resistance of a soil or rock to shearing stresses.

Specific gravity of solids—Ratio of the density of the solids (ρ_s) to the density of water (ρ_w).

Total, moist, or wet density—The total mass of partially saturated or saturated soil or rock per unit total volume.

Total, moist, or wet unit weight—The total mass of partially saturated or saturated soil or rock per unit total volume multiplied by standard acceleration of gravity.

Unit weight—The density multiplied by standard acceleration of gravity.

Void ratio—In soils and rock, the ratio of: (1) the volume of voids, to (2) the volume of solids in a unit total volume of soil or rock.

Water content—The ratio of the mass of water contained in the pore spaces of soil or rock material, to the solid mass of particles in that material, expressed as a percentage

6.2 CLAYS AND FINE-GRAINED SEDIMENTS

6.2.1 Introduction

Clays are composed typically of sheet-like particles because of their atomic structure (<0.002 mm). In contrast, silt (0.002–0.08 mm) and sand particles (0.08–2.0 mm) are the result of mechanical weathering. As an example, grey marine soft silty clays generally lay beneath the low, flat deltaic plains of South East Asia (SEA) to depths ranging from 10 to 30 m along the coastline to zero thickness away from the coast. The deltaic plains cover a large proposition of the total area of South East Asia as

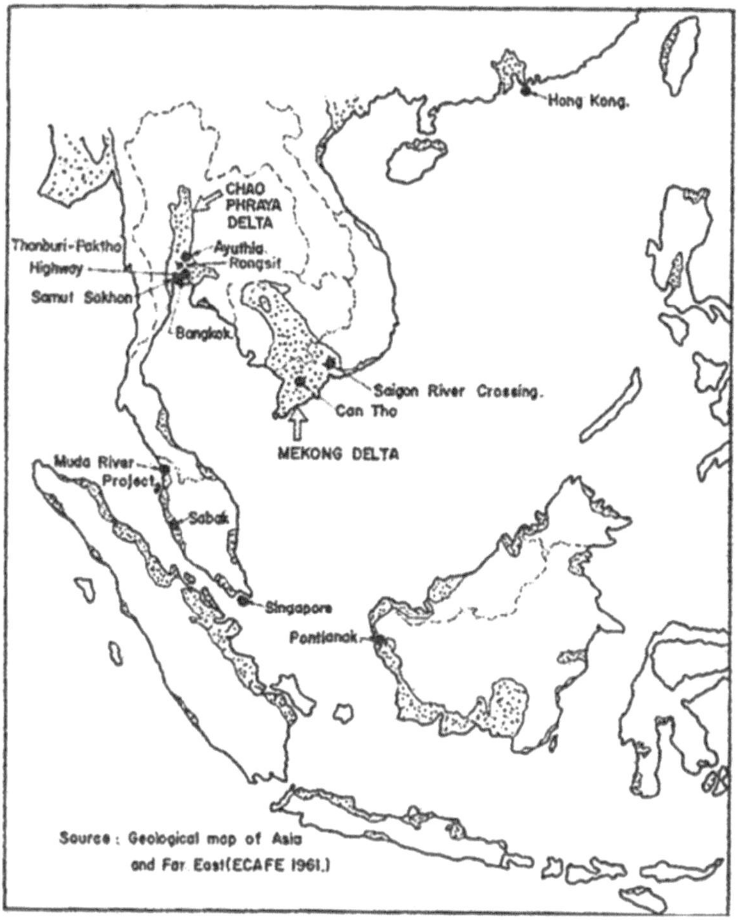

FIGURE 6.1 Distribution of recent clays in South East Asia (Cox, 1970).

shown in Figure 6.1. The deltaic plains are indicated in Figure 6.1 by the stippled areas. These clays are derived from material brought down by the major rivers in the area. The areal extent of these clays is greatest near the river mouths. The rate of rise in sea level generally exceeded the deposition rate of sediment over the last 20,000 years. This transgression of the sea over the land occurred until approximately 4000–8000 years ago. After this point there was little rise in sea level. During this period sediments were deposited under water (i.e., marine clays) as shown in Figure 6.2. After this period transgression sediments were deposited in extensive areas of coastal plains and deltas. These areas extend in widths up to 100 km. The marine clays have been subjected to wetting and drying cycles during periods of emergence. Additional buildup has occurred by terrestrial deposition of silt and clay size particles by annual flood waters. During this period the sediments have experienced secondary consolidation so that the majority of the plains area is only 0.5–3.0 m above mean sea level.

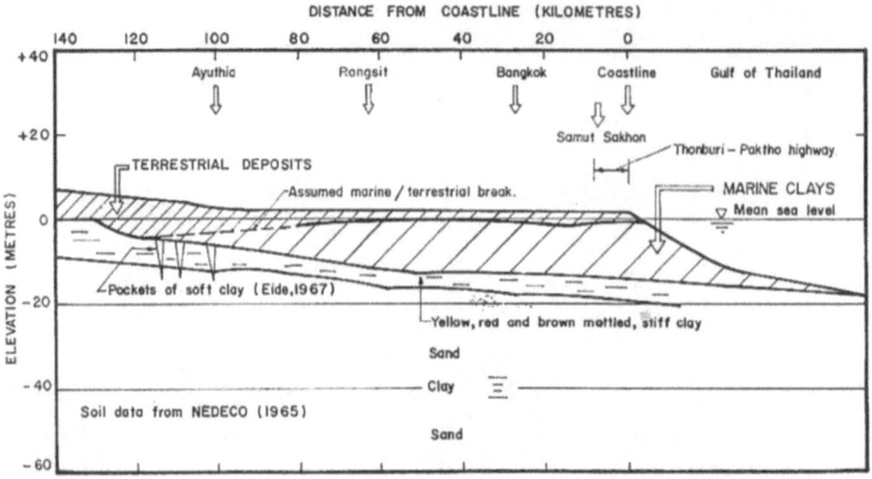

FIGURE 6.2 Illustration of cross section of coastline in the Gulf of Thailand (Cox, 1970).

6.2.2 Density and Water Content

The dry and wet density of a sediment is a function of the material's following properties: specific gravity of solids (G_s); its degree of compactness as represented by its void ratio (e); and its water content (w). The water content is defined as the ratio of the mass of water to mass of solids in a given volume at a standardized temperature.

6.2.3 Consistency

The clay percentage (<0.002 mm) of a marine soil in the area of the South China Sea in general varies from 35% to 60%, the silt percentage generally varies from 40% to 60%, and the sand percentage is generally less than 10% (Cox, 1970). The water content values range from approximately 60% to 100%. In areas having high organic contents between 2% and 5% the values of water content can be increased from 100% to 150%. The marine clays are highly plastic with liquid limits ranging from 50% to 150%. The corresponding Atterberg limit values generally plot close to the Casagrande A-line, indicating a slight organic content and a classification of CH/OH in the united system. In addition, sediments are largely over-consolidated ($I_L < 1$), Figure 6.3.

Clay contents of soils affect the basic index properties and geotechnical characteristics of the soils. A correlation between the clay content and (a) liquid limit w_L, (b) plastic limit w_p, and the (c) plasticity index I_p is presented in Figure 6.4. A review shows that a linear relationship exists between all three index parameters as a function of percent clay content. Based on results presented in Figure 6.4 the correlated equations are the following:

$$w_L = 1.70C + 13.5 \quad (6.1)$$

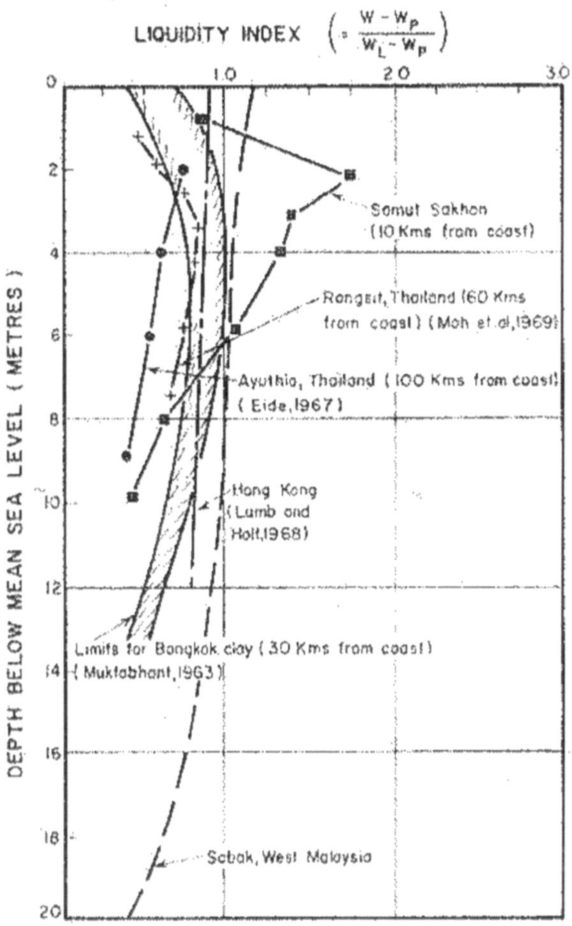

FIGURE 6.3 Liquidity index profiles for recent marine clays in SEA (Cox, 1970).

$$I_P = 1.26C \tag{6.2}$$

$$w_P = 0.44C + 13.5 \tag{6.3}$$

where C is the percent clay content.

Attwooll et al. (1985) reported that a marine clay and clayey silt at a dock at Tanjong Priok, Jakarta, Indonesia, was normally consolidated (NC) near the surface becoming progressively under consolidated (UC) with depth. The clay was very soft with a PI averaging 40% (ranged from 20% to 50%).

The clay particles generally have a normal activity of between 0.75 and 1.25. In contrast, inactive clays are often found adjacent to the coast where weathering of the clay particles has not begun. Breakdown of the clay minerals begins in the organic acid environment of the marine clays (pH 4.5–6.0). This breakdown of the clay

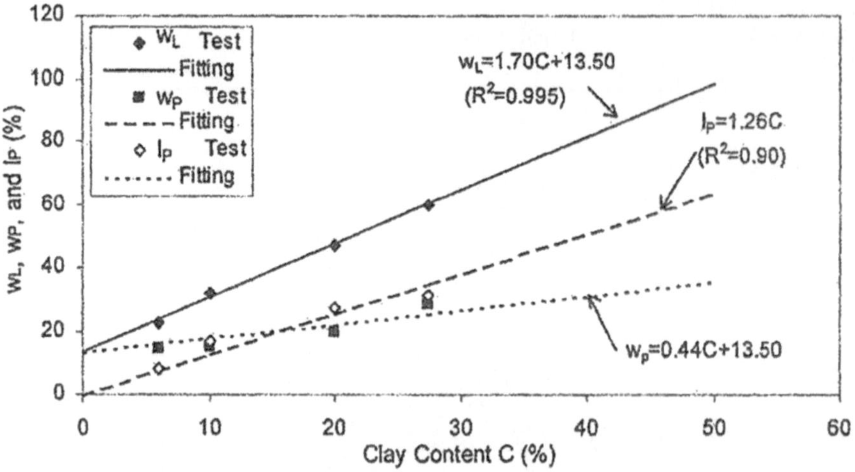

FIGURE 6.4 Measured data and correlations (by best fitting lines) between clay content and (a) liquid limit w_L, (b) plastic limit w_P, and (c) plasticity index I_P (after Yin et al. 2003). Reprinted with permission of Taylor & Francis.

increases rapidly when subject to oxygen-charged leaching waters, since this produces sulfuric acid and lowers the pH even further.

6.2.4 Compressibility

Consolidation tests have been carried out on clay samples from South East Asia using Casagrande type oedometers. Correlations of compressibility parameters with plasticity index are shown in Figures 6.5 and 6.6

Correlations of compressibility parameters and the plasticity index I_p by Yin et al. 2003 are as follows:

$$\text{Compression index} \quad C_c = 0.0138 I_p + 0.00732 \tag{6.4}$$

$$\text{Recompression or (unloading/reloading)} \quad C_r = 0.00219 I_p - 0.0104 \tag{6.5}$$

$$\text{Coefficient of secondary consolidation} \quad C_\alpha = 0.000369 I_p - 0.00055 \tag{6.6}$$

where

$C_c = -\Delta e / \Delta \log(\sigma_v^{'})$
$C_r = -\Delta e / \Delta \log(\sigma_v^{'})$
$C\alpha = -\Delta e / \Delta \log t$
Δe is the change in void ratio
$\sigma_v^{'}$ is the vertical effective stress
t is the creep time

The relationship between the coefficient of consolidation C_V versus I_P is shown in Figure 6.5. A review of Figure 6.5 shows that the relationship is non-linear.

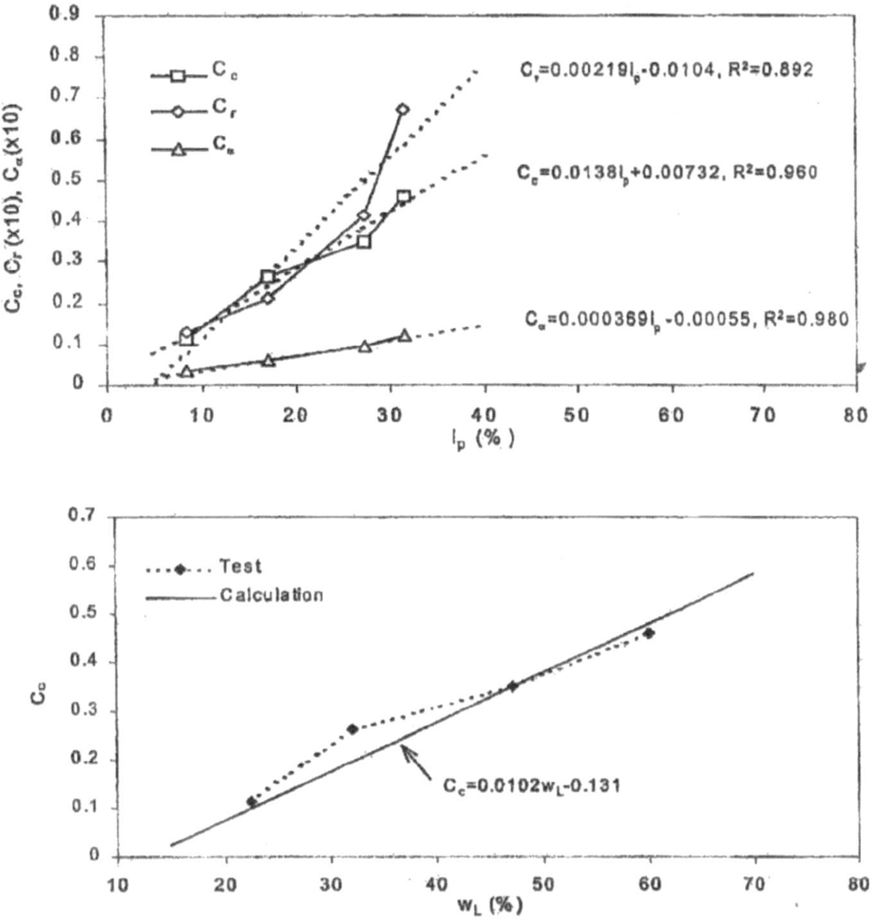

FIGURE 6.5 (a) Measured data and fitted lines of I_p vs. C_c, C_r, C_α for marine deposits (MD), and (b) comparison of measured data to calculated line for w_L vs C_c for MD (Yin et al., 2003). Note: Values of C_r and C_α in Figure 6.5a are multiplied by a factor of 10 to plot them on the same figure. Reprinted with permission of Taylor & Francis.

The correlations given above are similar to correlations reported by Nakase et al. (1988) for Kawasaki clay for both artificial mixtures and remolded natural marine clay as presented below.

$$C_c = 0.0104 I_P + 0.046 \tag{6.7}$$

$$C_r = 0.00194 I_P - 0.00892 \tag{6.8}$$

$$C_\alpha = 0.00033 I_P + 0.00168 \tag{6.9}$$

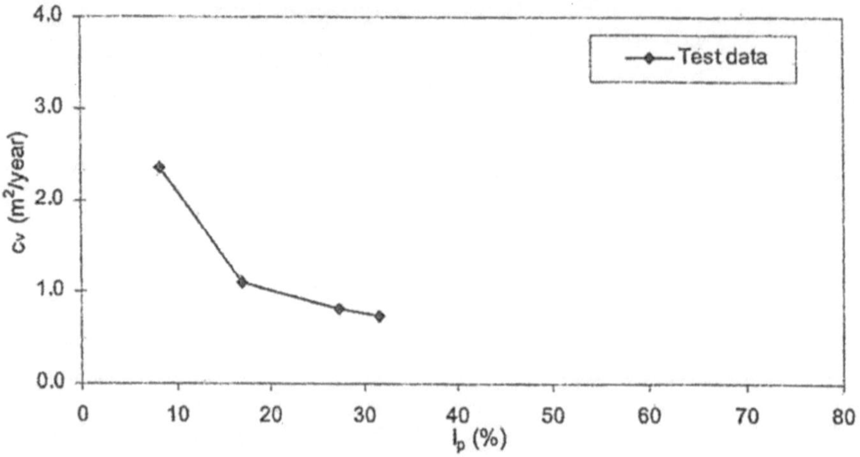

FIGURE 6.6 Coefficient of consolidation C_V versus Plastic Index I_p. Reprinted with permission of Taylor & Francis.

An empirical correlation relating the ratio $C_c/(1 + e_o)$ of the virgin consolidation line to the percentage water content of the natural soil (w) was proposed by Cox (1970). This correlation is given by the following:

$$C_c/(1+e_o) = 0.0045w \tag{6.10}$$

where

C_c is the compression index
e_o is the initial void ratio

Casagrande preconsolidation pressures have been determined from consolidation tests on a number of clays below the crustal zone in South East Asia as shown in Figure 6.7. The critical pressures (σ'_{vc}) generally exceed the overburden pressure (σ'_{vo}) and approximate the same value of $\sigma'_{vc} = 1.6\sigma'_{vo}$ given by Bjerrum (1967) for Drammen clay in Norway.

In Figure 6.7 it is shown that the marine clays are over-consolidated below the weathered zone. This effect could be due to a fall in sea level since deposition (Cox, 1970). The most likely evidence is an apparent over-consolidation effect, which is probably the result of secondary consolidation (Chaney and Almagor, 2016). A summary of apparent over-consolidation behavior in marine sediments is present in Table 6.2.

Secondary consolidation of the marine clays in South East Asia should be evident because they are highly plastic and possess relatively high organic contents. This reduction in void ratio (e) with time can be expected due to the decrease in liquidity index with increasing distance from the coastline.

A plot of the consolidation characteristics of Atlantic red clay, $e = 3.808$ with a compression index (C_c) of 1.0 is shown in Figure 6.8 for comparison.

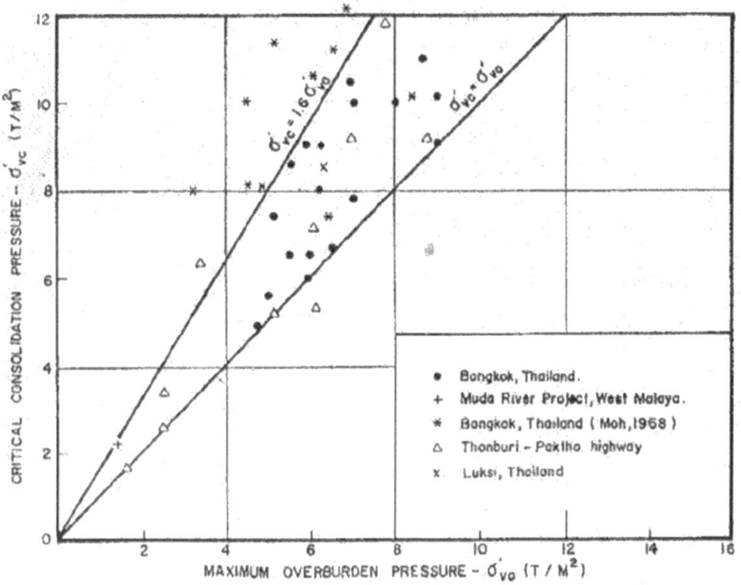

FIGURE 6.7 Critical pressures for recent marine clays in South East Asia (Cox, 1970).

6.2.5 Permeability

The void spaces between soil grains allow liquids to flow through them. A measure of this ability of a soil to transmit liquids is its permeability. Factors affecting permeability are (1) the effective grain size or effective pore size, (2) shapes of voids and flow paths through the soil pores—tortuosity, (3) saturation, and (4) viscosity of permeant. Darcy (1856) proposed an equation for calculating the velocity of flow of water through a sediment. This equation was originally developed for sandy materials, which assumes a linear relationship between the average surficial velocity of the fluid and the hydraulic gradient, with the proportionality constant or coefficient of permeability, k, being unique for a given sediment at a particular porosity, temperature, and pressure. Other, more complex, relationships have been suggested in attempts to account for actual flow conditions through fine-grained soils (Michaels and Lin, 1954; Olsen, 1965; Mitchell, 1976). A number of these relations depend upon various functions involving void ratio (refer to Figure 6.9). This Darcy relationship for velocity is presented in the following equation:

$$v = ki \tag{6.11}$$

where:

v is the velocity
k is the coefficient of permeability
i is the hydraulic gradient (dimensionless)

TABLE 6.2
Summary of Apparent Overconsolidation Behavior in Marine Sediments

No.	Area	Specific Location	Material Type	Sedimentation Rate (mm/1000 year)	Depth of Apparent Overconsolidation (M)	Underconsolidated	Normally Consolidated OCR = 1	Reference	Comments
1	Northwestern Pacific	East Bermuda Rise	Clay	300–400	17	–	X	Silva (1979)	The area has a sequence of rapidly deposited hemipelagic clays of thickness exceeding 100 m, water depth 4800–5500 m
2		Eastern SOHM Abyssal Plain	Clay		2.5	X	–	Silva and Jordan (1983)	
3		North Bermuda Rise Plateau	Clay		3–4	X	–	Silva and Jordan (1983)	
4	South Atlantic	Walvis Ridge	Foram nanno-fossil marl and ooze	41.0	10	Varies	Varies	Geotechnical Consortium (1984)	

5	North Pacific	North Central	Clay	0.3–4.5	4	–	X	Silva and Jordan (1983)	Abyssal hill region with sediment thickness 15–45 m. OCR attributed to very high interparticle bonding and possibly cementation by iron oxide, water depth 5800 m
6		DSDP Site 576A	Clay	10	20	–	X	Geotechnical Consortium (1984)	
7	Other areas	Oglofjorden and Dramsfjorden	Silt	Unknown	5	–	X	Richards (1976)	
8		Beaufort Sea	Silt	1.0	21.3	–	X	Wang and Vivatrat (1982)	

Source: Chaney and Almagor (2016). Reprinted with permission of Taylor & Francis.

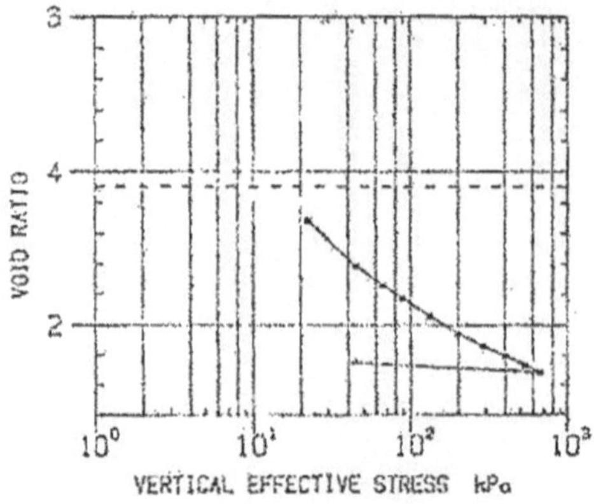

FIGURE 6.8 Consolidation characteristics, Atlantic red clay. $e = 3.808$, OCR = 4.2, $C_c = 1.0$, Institute of Oceanographic Sciences, UK.

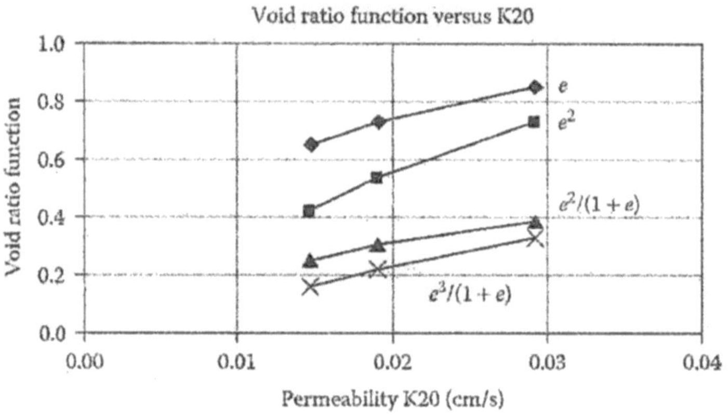

FIGURE 6.9 Permeability versus various ratio functions for a rounded sand (Chaney and Almagor, 2016). Reprinted with permission of Taylor & Francis.

The hydraulic gradient is defined as follows:

$$i = \Delta h / L \tag{6.12}$$

where

Δh is the piezometric head difference
L is the distance over which the Δh occurs

Water migration due to hydraulic gradients in marine sediments can be induced by natural processes such as sediment accumulation proposed by Gibson (1958) or by placement of structures on or in the seafloor. Water movement through clays is a very complex phenomenon because of the physicochemical activity of the particles and the effect of the surrounding double layers on the clay particles. Silva et al. (1981)

reported results of low-gradient permeability testing of fine-grained marine sediments using a modified back-pressured consolidation system for direct permeability measurements. They ran their experiments by first consolidating a sample under an effective stress. Once an equilibrium state was reached, a constant head permeability test was then conducted. These tests were conducted at varying gradients to study the validity of Darcy's equation and the occurrence of a threshold gradient. They found that most undisturbed samples (illite and smectite clays) did not show evidence of a threshold gradient. A threshold gradient was shown to become more probable as the sediment becomes denser. They did find that some artificially sedimented samples of illitic clay indicated threshold gradients of less than 5.

Many of the effects attributed to the presence of a threshold gradient can be attributed to the presence of undetected experimental errors. These errors are normally due to the following: contamination of measuring systems (Olsen, 1965), local soil consolidation and swelling, and bacterial growth (Gupta and Swartzendruber, 1962). A number of investigators using rigorous experimental procedures have not shown the presence of a threshold gradient (Gray and Mitchell, 1967; Mitchell and Younger, 1967; Miller et al., 1969; Olsen, 1969; Chan and Kenney, 1973). Mitchell (1993) did indicate that apparent deviations from Darcy's law could result from particle migrations leading to void plugging and unplugging, electrokinetic effects, and chemical concentration gradients.

Results of tests on five different types of deep-sea sediment are shown in Figure 6.10. These tests were performed on a combination of undisturbed and

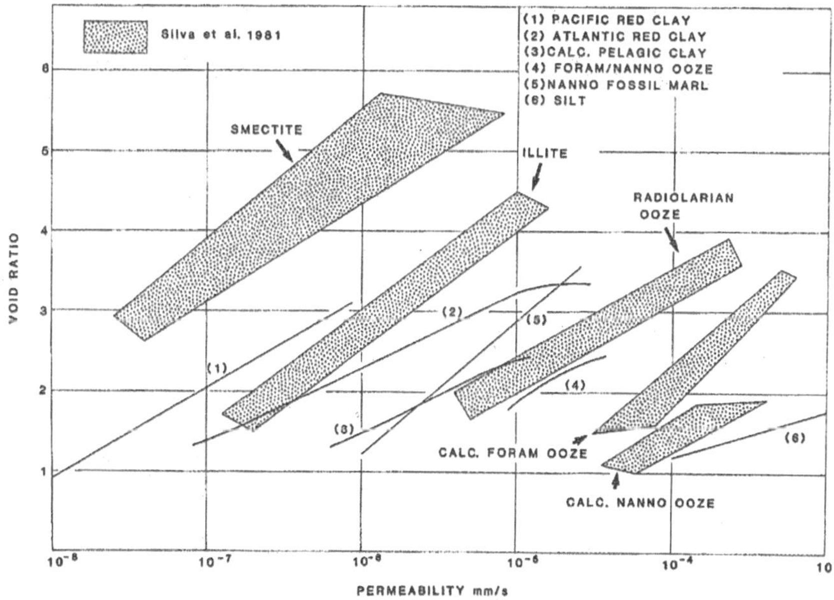

FIGURE 6.10 Permeability test results on different sediment samples (zones—Silva et al., 1981; and small lines—Schultheiss and Gunn, 1985). Institute of Oceanographic Sciences, UK.

TABLE 6.3
Permeability of Sediments from Abyssal Plains and Selected Regions

Location	Sediment Type	Coefficient of Permeability (cm/s)	Reference
Madeira Abyssal Plain (MAP)	Fine-grained turbidites	6×10^{-4}	Schultheiss and Gunn (1985)
MAP	Sandy and silty turbidites	$<1 \times 10^{-3}$	Schultheiss and Gunn (1985)
MAP	Pelagic ooze	1×10^{-6}	Schultheiss and Gunn (1985)
MAP	Pelagic clay	5×10^{-7}	Schultheiss and Gunn (1985)
MAP: Great Meteor East	Nannofossil turbidites (top)	1×10^{-4}	Schultheiss and Gunn (1985)
MAP: Great Meteor East	Silty turbidites (bottom)	5×10^{-4}	Schultheiss and Gunn (1985)
Nares Abyssal Plain	Fine-grained turbidites	1×10^{-7} to 1×10^{-8}	Shepard (1986}
N ares Abyssal Plain	Silty turbidites	1×10^{4} to 1×10^{-5}	Shepard (1986)
Kings Though, E. North Atlantic	Marls and ooze.	5×10^{-6}	Schultheiss and Gunn (1985)
Northwest Atlantic	Pelagic clay	8×10^{7} to 1×10^{-8}	Clukey and Silva (1982)
Gulf of Mexico	Various sediments	1×10^{-6} to 1×10^{-10}	Bryant et al. (1975)
Northwest Pacific	Pelagic clay (near seabed surface)	1×10^{-4}	Geotechnical Consortium (1985)
Northwest Pacific	Pelagic clay (50 m burial depth)	1×10^{-8}	Geotechnical Consortium (1985)

Source: Chaney, R. and Fang, H.Y. (1986) "Static and dynamic properties of marine sediments: A state of the art," ln R.C. Chaney and H.Y. Fang (eds.), *Marine Ceotechnogy and Nearshore/Offshore Structures.* STP 923, ASTM Press, Philadelphia, PA, pp. 74–11. Reprinted with permission, Copyright ASTM, 1986; also in Chaney; R/C et al., Abyssal plains: Potential sites for nuclear waste disposal. *Proceedings of the International Symposium on Environmental Geotechnology,* Allentown, PA, 1986.

artificially sedimented marine sediments. A review of this figure shows that the range of permeability is a function of material type and void ratio. The permeability of sediments from abyssal plains and selected regions is presented in Table 6.3 (Chaney and Fang, 1986).

The permeability for a variety of sediment samples have been plotted for comparison in Figure 6.10. These sediments are the following: (1) Pacific red clay, (2) Atlantic red clay, (3) calcareous pelagic clay, (4) foram nanno ooze, (5) nanno fossil marl (turbidite), and (6) silt base of turbidite. A review shows that the permeability void ratio relationships for these six sediment types varies considerably. At the end of the spectrum the red clays with their very fine-grain sizes have the lowest permeability with the Pacific red clay being less permeable than the Atlantic red clay for the same void ratio. Because of the large compressibility index (C_c) and the high initial void ratio (e) the permeability changes very rapidly in the first 50 m of burial.

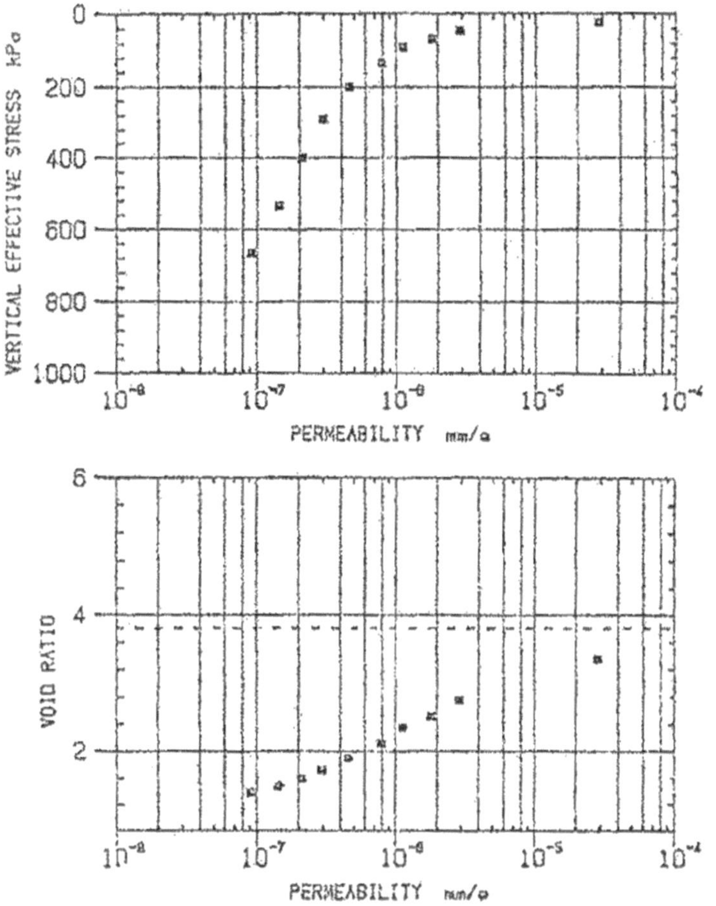

FIGURE 6.11 Permeability characteristics, Atlantic red clay, e = 3.808, OCR = 4.2, Schultheiss and Gunn (1985). Institute of Oceanographic Sciences, UK.

In comparison, the calcareous pelagic clay, the nannofossil turbidite, and the foram nanno marl increase in permeability in that order. The silty base of the turbidite has a permeability greater than some 100,000 times more than the red clay. A comparison between data presented by Schultheiss and Gunn (1985) with permeability data from Silva et al. (1981) can be seen in Figure 6.10. It can be seen that while the samples do not cover the higher void ratios, the overall pattern of the data is very similar. As an example, permeability of Atlantic red clay is presented in Figure 6.11.

6.2.6 Shear Strength

Idealized shear strength profiles for homogenous marine deposits that are (1) normally consolidated (NC) clay, (2) under-consolidated (UC) clay, and (3) over-consolidated (OC) clay are presented in Figure 6.12. A review of Figure 6.12a shows

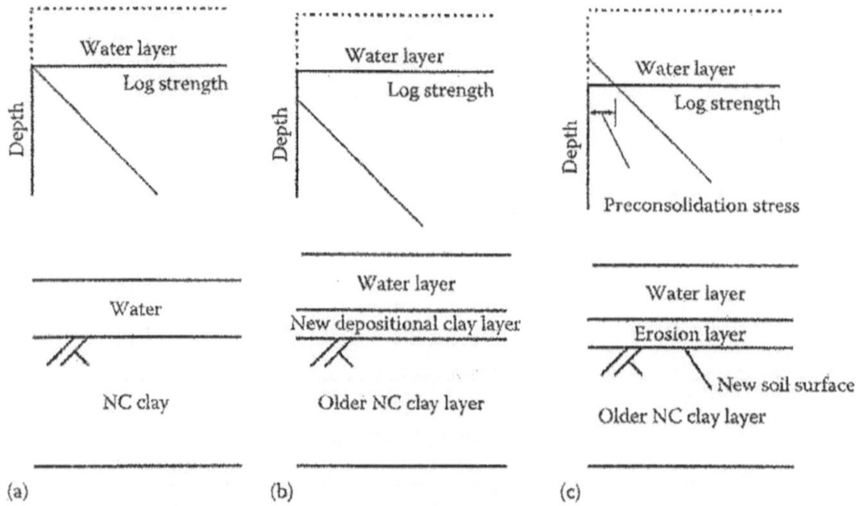

FIGURE 6.12 Idealized strength profile in homogeneous clay. (a) NC strength profile; (b) UC strength profile; (c) OC strength profile.

that the NC clay profile exhibits a linear shear strength behavior starting at zero at the water/sediment interface and increasing with depth. Figure 6.12b depicts an UC clay profile where a new clay layer overlays an older clay layer that has not reached equilibrium under its previous loading. This new clay layer results in an increase in excess pore pressure in the older layer. The resulting shear strength in the older clay layer is therefore decreased. For an OC clay (Figure 6.12c), the clay layer has experienced a larger load in its geologic past than at present exists. The clay layer remembers this past loading in the form of a cohesion and the resulting shear strength versus depth is increased. The strength of sediments in the above is a function of its structure (i.e., assemblage of grains), moisture content, and the properties of the mineral grains forming the solid phase of the system. The strength and shearing resistance of soils (sediments) are the most important characteristics for engineering design or consideration of hazards.

The effect of plasticity index on the ratio of shear strength (C_u) to the effective overburden stress (σ'_{vo}) for some marine clays of South East Asia is presented in Figure 6.13. A review of this figure indicates that the shear strength increase parameter $\frac{C_u}{\sigma'_{vo}}$ is approximate to Skempton's correlation at sites adjacent to the present coastline, where the primary consolidation is complete, but increases for sites away from the coastline, where the sediments are much older and have undergone secondary consolidation after deposition.

As an example, the stratigraphy of an off shore site by Jakarta, Indonesia, consists of 6 units. Unit 1 consists of marine clay and clayey silts from 6 to 25 m. Overlying this is a 4–6 m layer of very soft clay. Unit 2 consists of tuffaceous silty sand and sand silt (25 to 35 m) very stiff and dense volcanic materials that are slightly cemented. Unit 3—35–75 m initially normally consolidated but becoming progressively

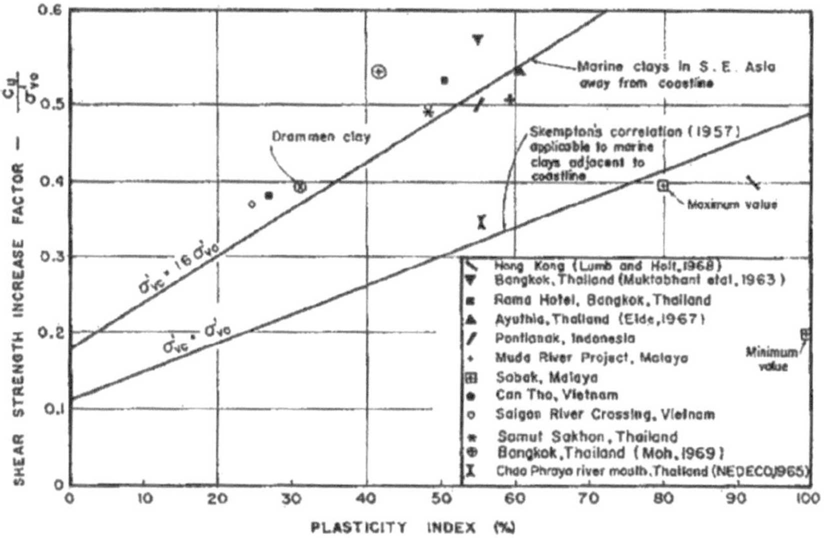

FIGURE 6.13 Undrained shear strength relationships for recent marine clays (Cox, 1970).

under-consolidated with depth because of artesian pressures from an underlying aquifer. The clays are stiff to very stiff. Unit 4—Sand 75–80 m. A dense sand, originally under high artesian pressure but falling because of groundwater development. Unit 5—80 to approximately 125 m clay similar to upper clay unit but highly underconsolidated. Unit 6 – Clayey sand – 125 to 130 m.

Strength testing offshore Jakarta, Indonesia, was conducted using an in situ vane shear, laboratory TXUU (triaxial unconsolidated undrained) and torvane. These strength tests were conducted in the upper 25 m. Results are presented in Figure 6.14. A review of Figure 6.14 indicates a wide variation in test results depending on the test method that was employed.

Shear strength profiles away from the coastline in South Vietnam are presented in Figure 6.15.

The shear strength increase factor $\frac{C_u}{\sigma'_{vo}}$ as a factor of PI is plotted in Figure 6.15.

A review of Figure 6.15 indicates that with the exception of results from the Chao Phraya River mouth and at Hong Kong, the majority of parameters consistently fall way above the correlation of Skempton (1957) for marine clays in the temperate climates of the northern hemisphere.

Sabak is located in western Malaya close to the mouth of the Surgai Bernam river, The shear strength profiles indicate that the sediments in this region could still be undergoing primary consolidation (Figure 6.16). A summary of angles of shearing resistance as a function of plasticity index for recent Southeast Asia marine clays is presented in Figure 6.17. A review shows that the angle of shearing resistance decreases with increasing plasticity index.

The effect of percent clay content and Plasticity Index on the effective friction angle ϕ' is shown in Figure 6.18. A review of Figure 6.17a shows that as clay content increases from 0% to 65% the angle of effective stress decreases from 32° to approximately 20°.

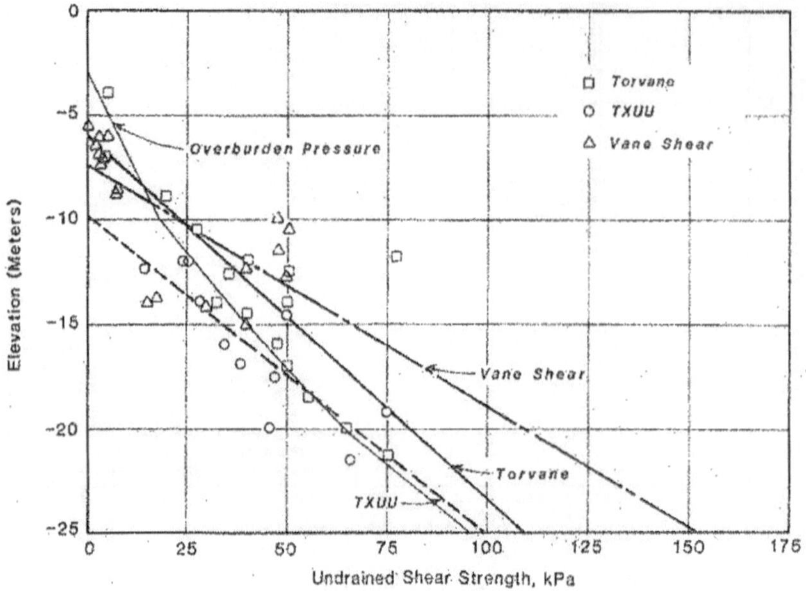

FIGURE 6.14 Jakarta, Indonesia, site shear strength test results from off-shore borings (Attwooll et al. 1985). Reprinted with permission of ASTM.

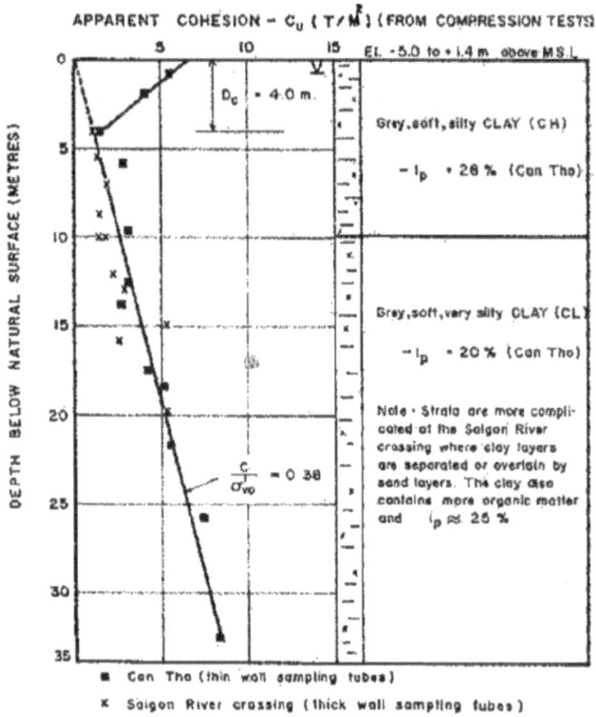

FIGURE 6.15 Shear strength profiles for recent clays in South Vietnam (Cox, 1970).

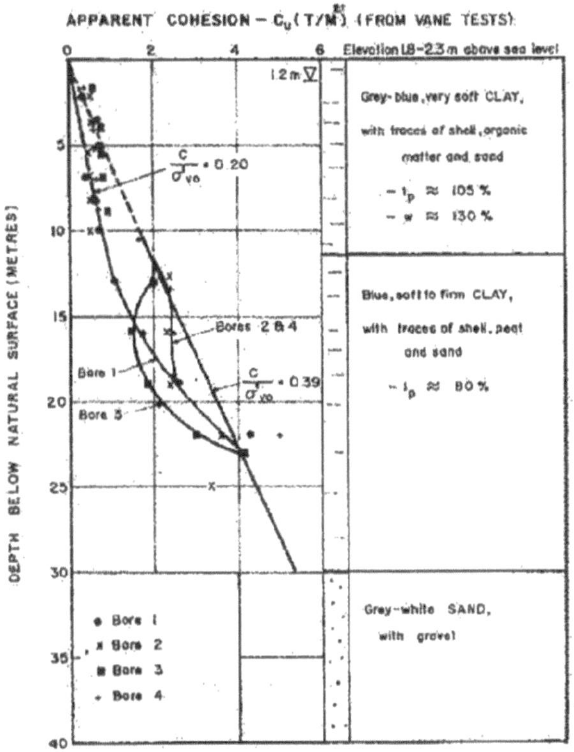

FIGURE 6.16 Shear strength profiles for recent clays at Sabak, Malaya (Cox, 1970).

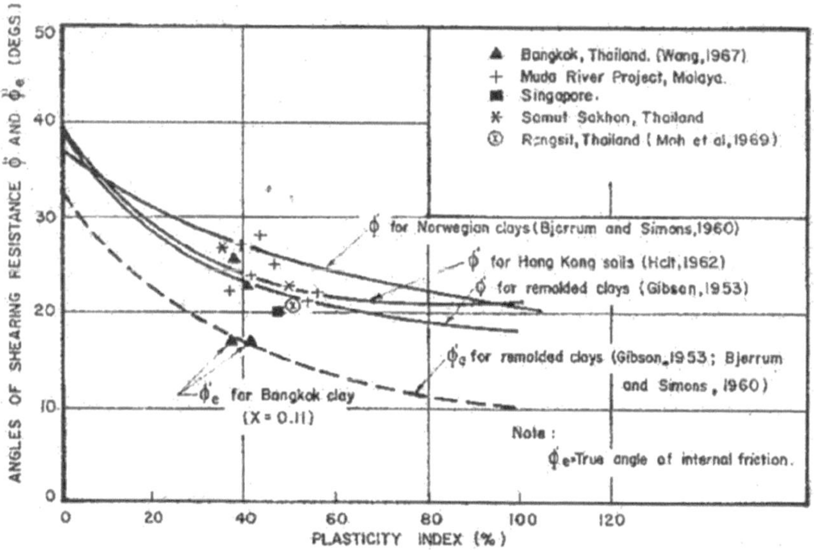

FIGURE 6.17 Effective stress angles of shearing resistance for recent marine clays in South East Asia (Cox, 1970).

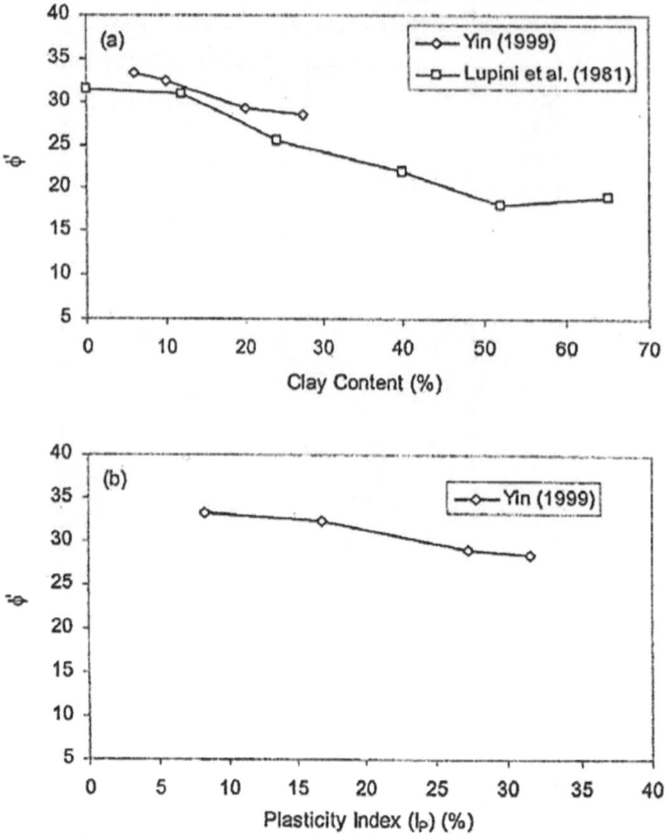

FIGURE 6.18 Effective friction angle φ′ vs. clay content, and (b) plasticity index I_p for the marine deposits (data from Yin 2003 and Lupini et al. 1981). Reprinted with permission of Taylor & Francis.

This change occurs in an approximately linear manner. A similar relationship occurs between the angle of effective stress and plasticity index as shown in Figure 6.18b.

Figure 6.19a shows the curves of the secant Young's modulus E vs clay content for consolidation pressures of E_{50}/σ'_{3c} = 100 kPa, 200 kPa, and 400 kPa. If the Young's modulus is normalized using the effective confining pressure σ'_{3c} then E_{50}/σ'_{3c} = 100 kPa, 200 kPa, and 400 kPa, the normalized value of E_{50}/σ'_{3c} versus clay content is linear as shown in Figure 6.19b. A straight line is used to fit the data in Figure 6.19b.

$$\frac{E_{50}}{\sigma'_{3c}} = -8.8C + 280. \tag{6.13}$$

Since it was shown earlier that $C = I_p/1.26$ then

$$\frac{E_{50}}{\sigma'_{3c}} = -6.98I_p + 280. \tag{6.14}$$

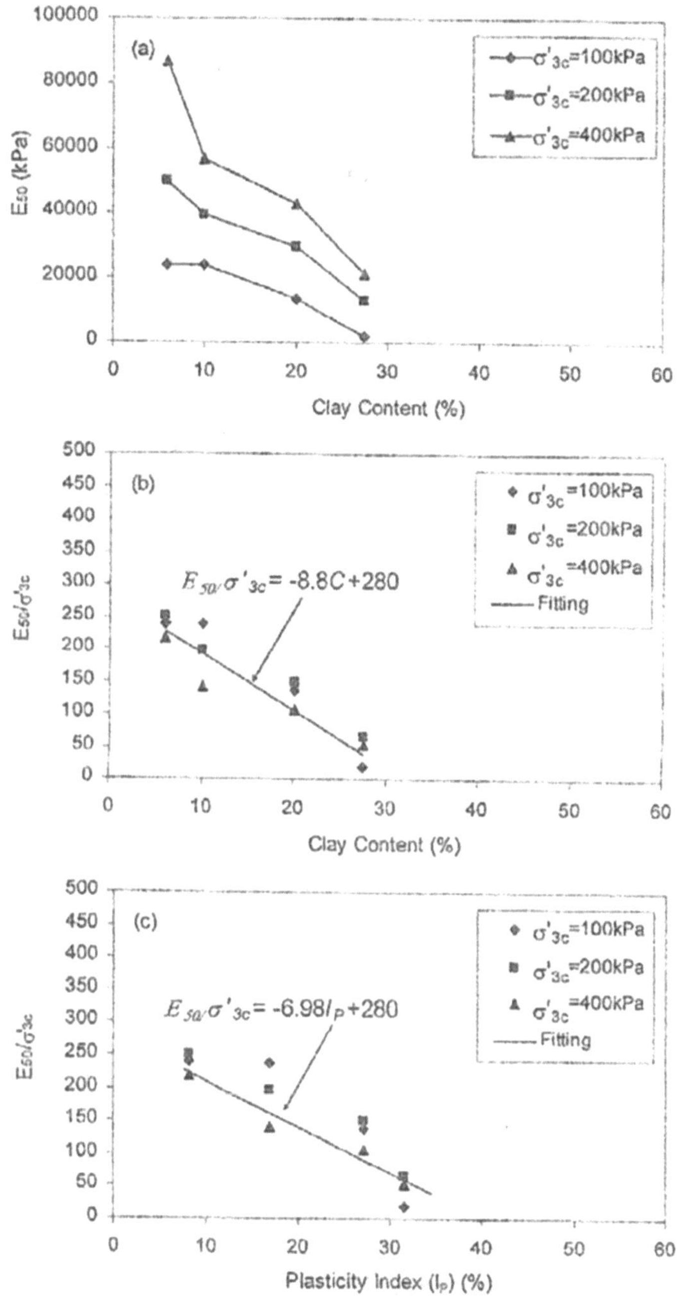

FIGURE 6.19 (a) Secant Young's modulus E_{50} vs. clay content, (b) normalized Young's modulus E_{50}/σ'_{3c} vs. clay content, and (c) normalized Young's modulus E_{50}/σ'_{3c} vs. plasticity I for marine deposits (Yin et al. 2003). Reprinted with permission of Taylor & Francis.

6.3 CALCAREOUS MATERIALS

6.3.1 Introduction

Seabed investigations have revealed that surficial sediments contain large quantities of calcium carbonate. These carbonate sediments are the skeletal remains of microscopic plants and animals that thrive on nutrients in surface waters but ultimately settle to the seafloor and supplement other non-carbonate forms of sediments. The primary classes of carbonate particles are pteropods, Foraminifea, or forams, and coccolithophorids, which are also called nannofossils or nannos because of their small size. As indicated above, calcareous materials come from a variety of origins, shapes, and sizes as illustrated in Figure 6.20. Sediment type is given by a broad definition based on the amount of carbonate and microscopic observations. The following broad classifications as presented in Tables 6.4 and 6.5 have been used.

6.3.2 Density and Water Content

Idealized foram and nannofossil cross sections of individual particles are presented in Figures 6.21a and 6.21b respectively. In these figures the intra-particle water is

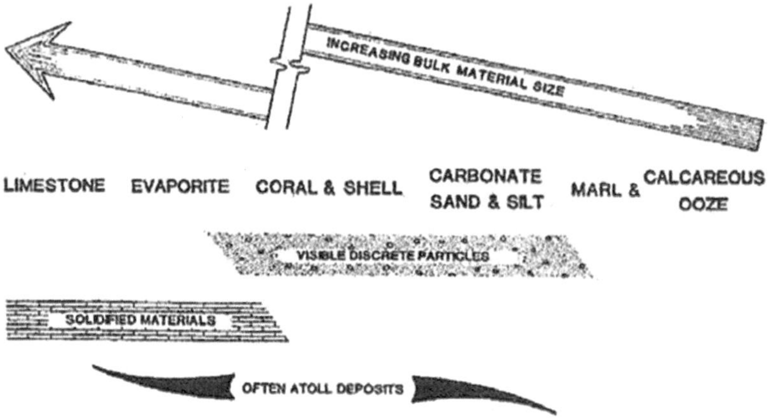

FIGURE 6.20 Schematic illustration of types of Calcareous materials (Angemeer and MCNeilian, 1982). Reprinted with permission of American Society for Testing and Materials.

TABLE 6.4
Broad Classification of Carbonates

Carbonate (%)	Sediment
100–70	Ooze
70–30	Marl
30–5	Calcareous clay
5–0	Clay

Source: Schultheiss and Gunn (1985) and IOS (1978).

TABLE 6.5
The Carbonate Fraction Is Typically Divided up into Percentages of Forams and Nannofossils

Forams (%)	Percent Nannofossils (%)	
100–70	0–30	Foram
70–30	30–70	Foram nanno
30–03	7–100	Nanno

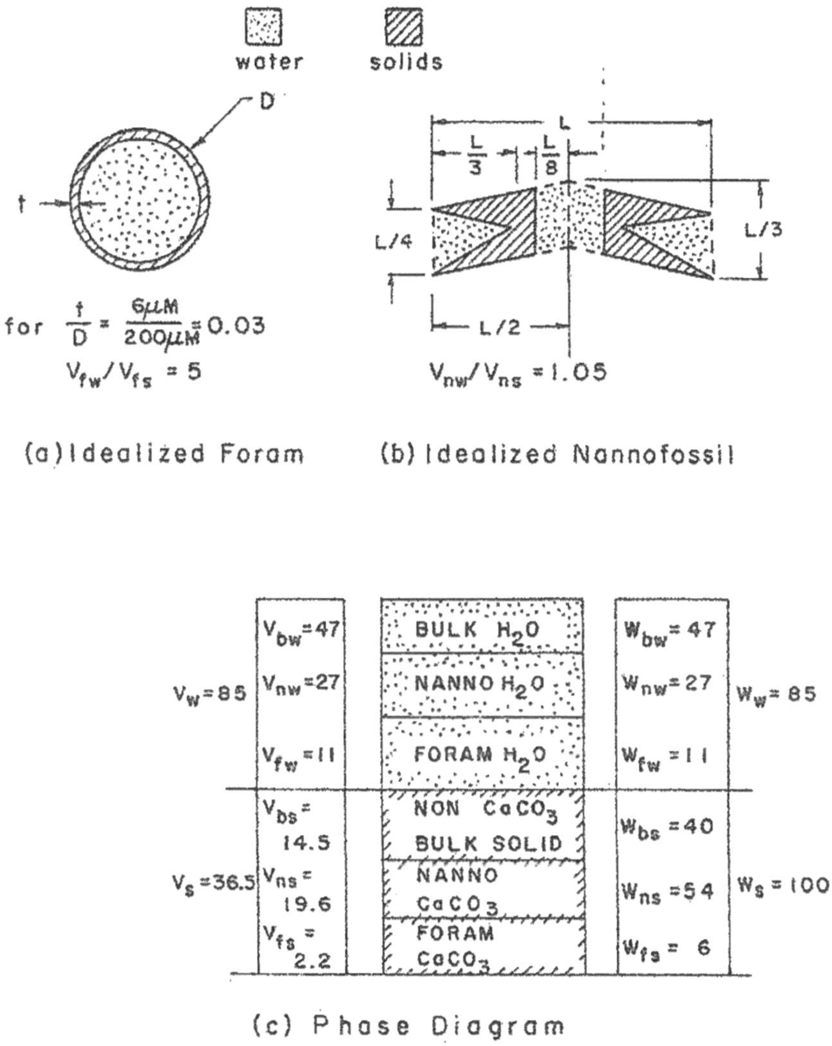

FIGURE 6.21 Estimate of weight–volume relationships for Core s-4 (Demars, 1982). Reprinted with permission of American Society for Testing and Materials.

shown stippled and the solid parts are shown as solid lines. The ratio of the intra-particle water volume to particle solid volume has been estimated for nannoforam ooze by Demars (1982), For an average foram idealized as a hollow sphere with an outside diameter of 200 μm and a wall thickness of 6 μm the volume ratio of water (V_{fw}) to solids (V_{fs}) is approximately 6.0. In contrast for a nannofossil the volume ratio of water (V_{fw}) to solid's (V_{fs}) is approximately 1.06. An estimate of weight volume relationships for a core from the Eastern Atlantic off North Africa at a depth of 4300 m is presented in Figure 6.21c.

Assuming the percent weight of foram solids (M_{fs}) is 6% in a sediment, nanno solids (M_{ns}) is 54% and non-carbonate bulk solids (M_{bs}) is 40%, a specific gravity of 2.75 and using volume ratios as given previously, the volumes and weight of solid and liquids can be computed as shown in Figure 6.21c. This method estimates that the total water content of 83% may be broken down into foram, nanno and interparticle water contents of 11%, 27%, and 47% respectively. Thus the interparticle water content of 47% can be increased to 85% if intra-particle water content is included. Thus even through this is a very simplified analysis it is evident that the quantities of intra-particle water are very large. This greatly affects water content of a sediment by the amount of forams that are present.

Lee (1982) *modeled* a calcium carbonate sediment as a bimodal fine grain matrix with carbonate inclusions in sediment as shown in Figure 6.22. In this model he considered the relation between carbonate content and density as consisting of two end points. These end points were the non-carbonate matrix sediment alone and carbonate at the other. In between these two end points were increasing carbonate content. Using this model Lee (1982) derived a relation for the variation of the total bulk density of the system (ρ_t) as a function of carbonate content (C), porosity of non-carbonate matrix (n_m), porosity of carbonate grains (n_c), non-carbonate grain density (ρ_{am}), and carbonate grain density (ρ_c) as shown below:

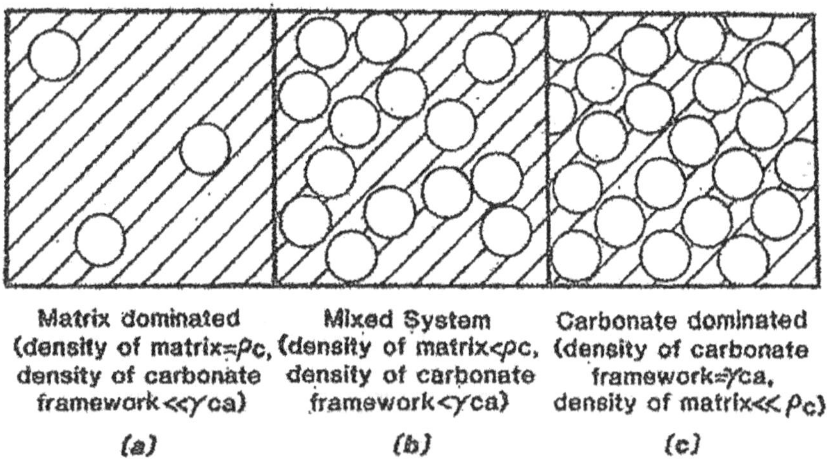

FIGURE 6.22 Three phases of a model for a bimodal carbonate, non-carbonate matrix system (Lee, 1982). Reprinted with permission of ASTM.

$$\rho = \frac{m_c - m_{ca}}{V_t} \tag{6.15}$$

$$\rho_t = \frac{\rho_{sm}(1-n_c)(\rho_{ca}-\rho_m)*C}{\rho_c(1-n_c)-\left[\rho_c(1-n_c)-\rho_{sm}(1-n_m)\right]C} * \rho_m \tag{6.16}$$

$$\rho_{t2} = \frac{\rho_c(\rho_{sm}-\rho_w)(1-C)(\rho_{ca}-\rho w)}{\rho_{sm}(\rho_c-\rho_w)C} + \rho_{ca} \tag{6.17}$$

These equations can be simplified by assuming $n_m = 0$, and $\rho_{am} = 2.72$ Mg/m

$$\rho_{t1} = \frac{\rho_{sm}(1-n_m)(2.72-\rho_m)C}{2.72-\left[2.72-\rho_{sm}(1-n_m)C\right]} + \rho_m \tag{6.18}$$

$$\rho_{t2} = \frac{2.72(\rho_{sm}-1.025)(1-C)(\rho_c-1.025)}{\rho_{sm}(1.695)C} + \rho_c \tag{6.19}$$

where

ρ_{t1} is the total bulk density when matrix material controls behavior
ρ_{t2} is the total bulk density when carbonate skeleton controls behavior
ρ_c is the carbonate grain density, (2.72 Mg/m^3)
ρ_{ca} is the bulk grain density of carbonate grains (including intrastitial voids not filled with matrix)
ρ_{tc} is the total bulk density of carbonate
ρ_m is the bulk density of non-carbonate matrix
ρ_{sm} is the non-carbonate matrix grain density
$n_c = \frac{\rho_{ca}-\rho_c}{\rho_w-\rho_c}$ is the inherent porosity of carbonate grains
$n_m = \frac{\rho_c-\rho_{sm}}{\rho_w-\rho_{sm}}$ is the porosity of non-carbonate grains
C is the carbonate content (percent dry weight)
V_t is the total volume
M_c is the mass of non-carbonate grains alone
M_{ca} is the mass of carbonate grains alone
ρ is the total density

The critical carbonate content (C_{cr}) roughly separates matrix and carbonate framework dominated behavior.

For $C < C_{cr}$ normalize as follows: C/C_{cr}
For $C > C_{cr}$ normalize as follows: $(1 + C - C_{cr})/(1 - C_{cr})$

The clay dominated matrix material (higher ρ_{sm}) is much less open in structure and, therefore, has a higher bulk density.

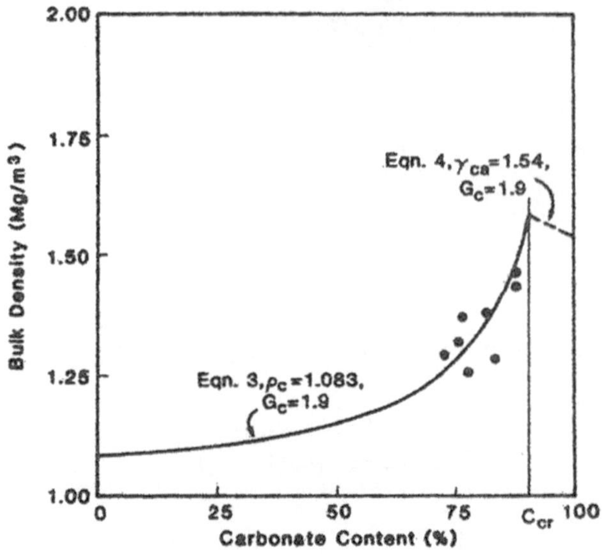

FIGURE 6.23 Bulk density versus carbonate content for the Eastern Equatorial Pacific (Lee, 1982). Reprinted with permission of ASTM.

Using Equations 6.4 and 6.5 it is possible to fit them to sets of bulk density–carbonate content data using appropriate values of ρ_c, and ρ_{sm}. This is *opposed to* calcareous oozes, which are made from skeletons of calcium carbonate organisms (i.e., coccolithophores). A figure of calculated and actual experimental data of bulk density versus carbonate content for the eastern equatorial pacific is presented in Figure 6.23. A review shows a close agreement.

6.3.2.1 Size Ranges of Particles

A sediment is composed of various sizes of particles as shown in Figure 6.24. The size of a particle is an important property of sediments. This is because size determines the mode of transport and distance it will travel before settling to the bottom.

If you are a geologist a sediment consisting of single-sized particles is considered well sorted. A well sorted sediment is one that has a limited size range of particles. The other sized particles in the sediment have been removed, usually by currents or waves. If you are an engineer it is just the reverse (i.e., poorly sorted). A poorly sorted sediment contains different sized particles. This usually indicates that little mechanical energy has acted to sort the particles.

6.3.3 Consistency

6.3.3.1 Effect of Increasing Calcium Carbonate Content

A sediment with increasing calcium carbonate content exhibits decreasing plasticity and behaves like a frictional material. Figure 6.25 illustrates the reduction in both liquid limit (w_L) and plasticity index (I_p) with increasing carbonate content. This

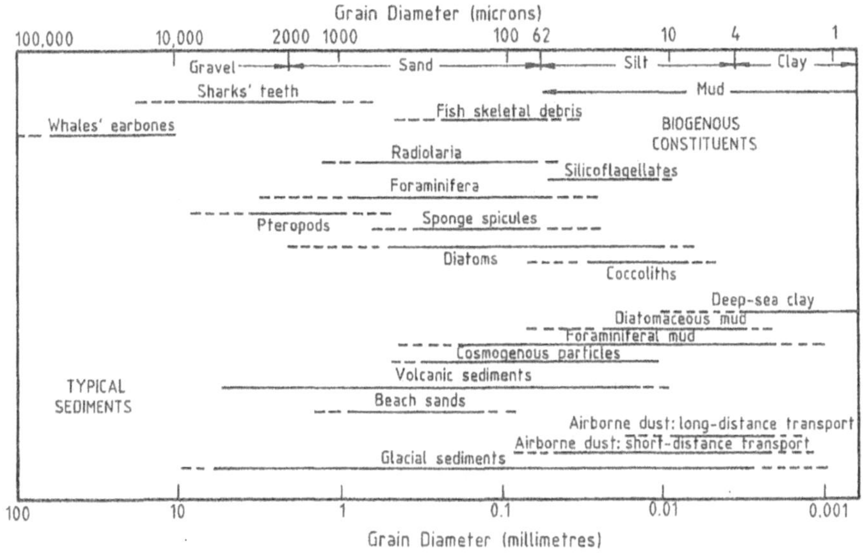

FIGURE 6.24 Typical sizes of particles from various sources (Gross, 1967). Public domain.

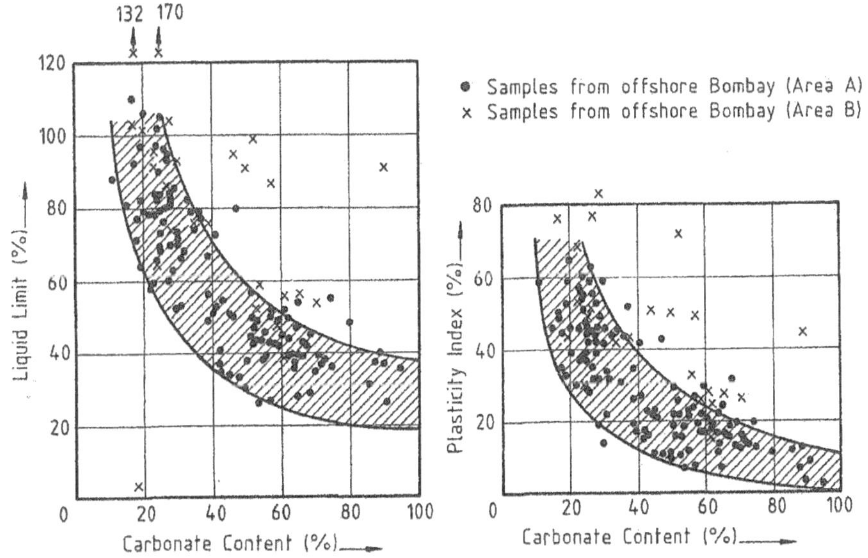

FIGURE 6.25 Carbonate content versus liquid limit and plasticity index (Beringen et al., 1982). Reprinted with permission of the American Society for Testing and Materials.

behavior occurs because unlike clay carbonate particles are not affected by surface forces (i.e., they have no surface electrical charges). Therefore they are not affected by the electrical polar characteristics of water.

6.3.3.2 Water Content Changes Due to Particle Crushing

Nacci et al. (1974) conducted consolidated undrained and drained triaxial tests on cemented carbonate sediments. At small strains (<0.5%) the cemented samples usually did not develop pore pressures but rather acted similar to a soft rock. At larger strains, cementation between particles began to break down and pore water pressures began to develop. After the cementation breakdown the carbonate soil tends to behave in a manner similar to an un-cemented calcareous silt. Under hydrostatic stress, cementation breakdown tends to occur at pressures of between approximately 1–5 MPa depending on the strength of the cementation bound (Poulos, 1988).

When classification tests are performed it is not known if particles exhibit breakage and water release. The quantity of intra-particle water that these sediments contain may be partly evaluated by a simple plot of water content versus carbonate content for each of the specimens as shown in Figure 6.26. It is observed that the fragmented or partially disintegrated specimens (i.e., S-1, S-2) have a much lower water content than that of the cores, which show no signs of particle dissolution and disintegration (i.e., S-3, S-4), Figure 6.26.

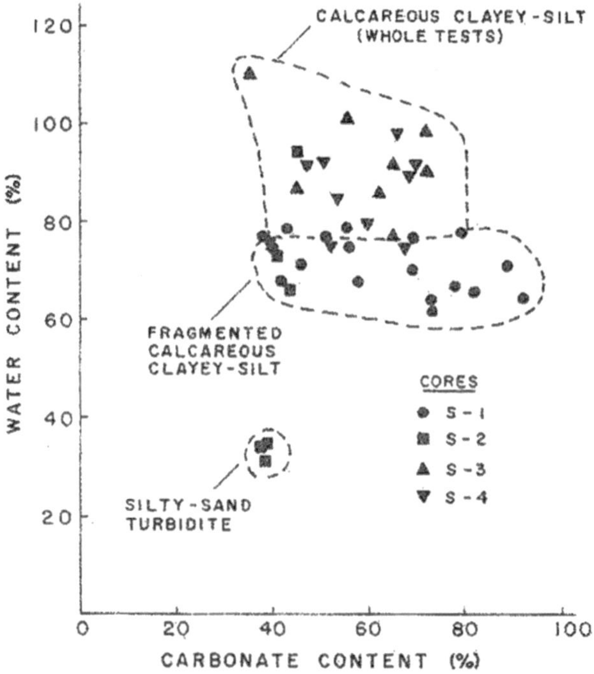

FIGURE 6.26 Influence of carbonate content on water contents for test specimens (Demars, 1982). Reprinted with permission of American Society for Testing and Materials.

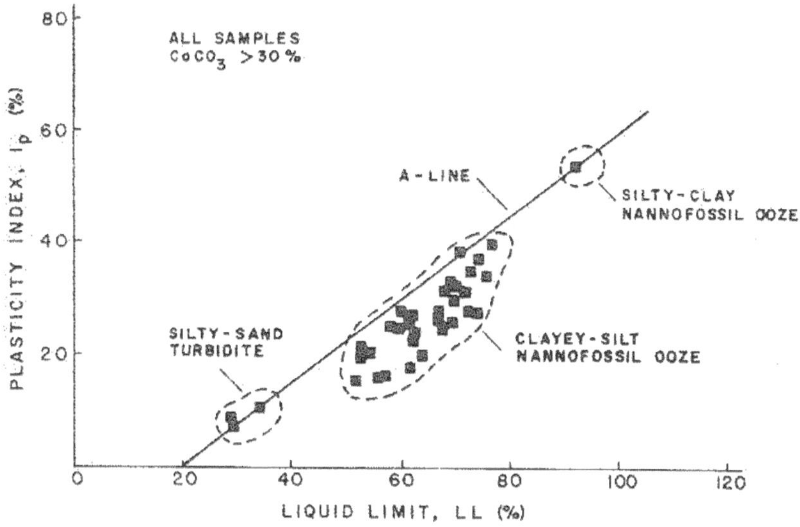

FIGURE 6.27 Plasticity chart for (1) silty-sand turbite, (2) clayey-silt nannofossil ooze, and (3) silty-clay nannofossils ooze (Demars, 1982). Reprinted with permission of American Society for Testing and Materials.

As an example, the above sediments are uniformly classified as silts or organic silts of high plasticity (MH) by USCS as shown in Figure 6.27. Because of this breakage the classification of carbonate soils may be in error. This error depends upon the quantity of intra-particle water associated with the calcareous nannoforam sediment released upon crushing. This water is included in determination of w_L and w_p but nullified in the $I_p = w_L - w_p$.

As an example, the data in Figure 6.27 should probably be shifted to the left by some unknown amount for purposes of classification based on using index properties.

This estimate is only approximate since all carbonate particles were not disintegrated. In addition, particles that were disintegrated increased packing and reduced intra-particle voids.

6.3.4 Compressibility

6.3.4.1 Primary Consolidation

Consolidation tests run indicate that plots of void ratio (e) as a function of the logarithm of effective stress is not always linear. In addition, at high stresses the compressibility decreases significantly. Bryant et al. (1974) has presented typical consolidation curves for calcareous soils in Figure 6.28. A review of Figure 6.28 shows that the curved shape is similar to what is observed in some sensitive clays and silica sands at high stress levels. The curved consolidation curve results in the compression index (C_c) being a function of the effective stress level. The C_c for calcareous soils is similar to clays because it is a function of the original void ratio (e_o).

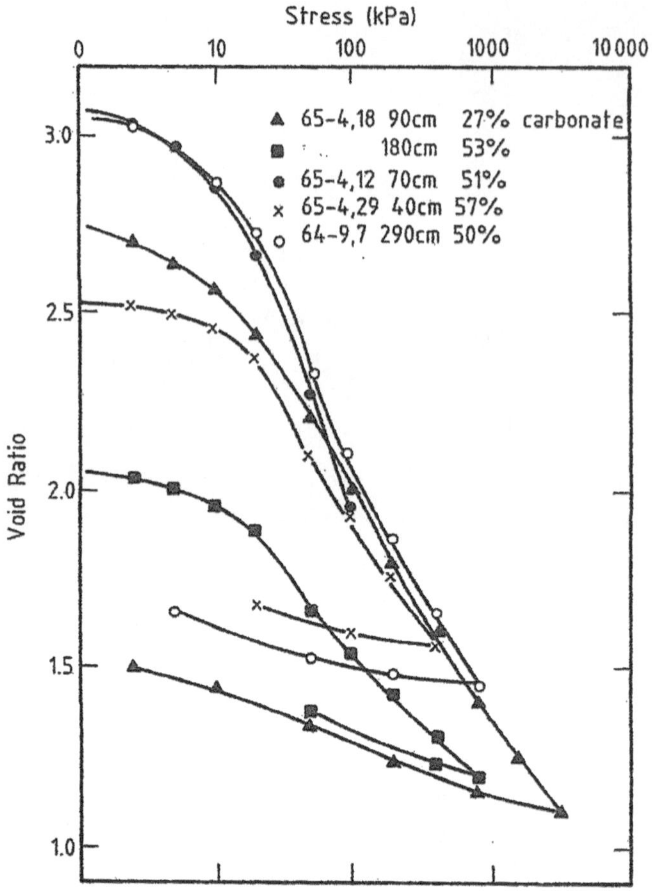

FIGURE 6.28 Typical consolidation curves for calcareous soils (Bryant et al., 1974). Reprinted with permission of Springer Nature.

Poulos (1988) has presented a summary of one-dimensional compressibility data for Bass Strait carbonate sand in Figure 6.29. In this figure both consolidometer and K_o consolidation tests data were presented. Higher values of C_c than those shown in Figure 6.29 have been presented in the literature (i.e., Nacci et al. 1974, Bryant et al. 1974, Nauroy and le Tirant 1981). Compression index C_c values as high as 0.9 have been recorded for soft normally consolidated deposits with initial void ratios of the order of 3. These values of C_c are similar to a linear extrapolation of the data in 6.29. Bryant et al. (1974) found that C_c decreases in general with increasing carbonate content, and this trend is consistent with the transition in behavior from cohesive to granular with increasing carbonate content (Demars et al. 1976). Valent et al. (1982) tested two calcareous oozes from the Caribbean as presented in Figures 6.30 and 6.31. Figure 6.30 indicates that a e versus log σ_v' curve does not become a straight line at higher pressures for a load increment ratio $\Delta\sigma_v' / \sigma_v' = 1$. A straight line e versus log σ_v' curve does occur when testing ooze with a load increment ratio

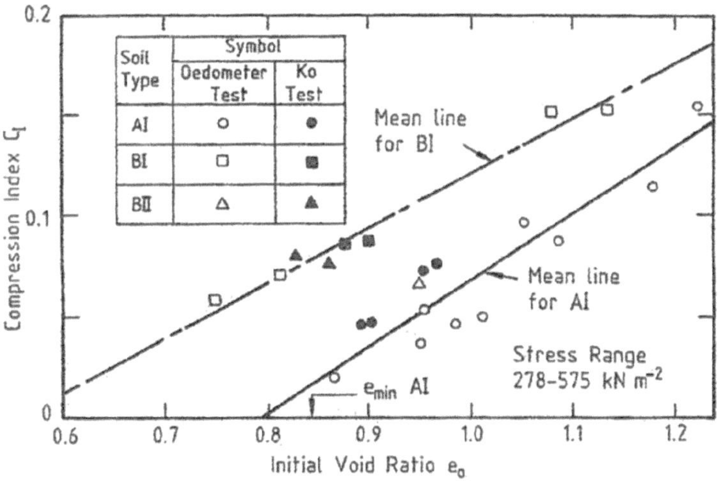

FIGURE 6.29 A summary of one-dimensional compressibility data for Bass Strait carbonate sand (Poulos, 1988).

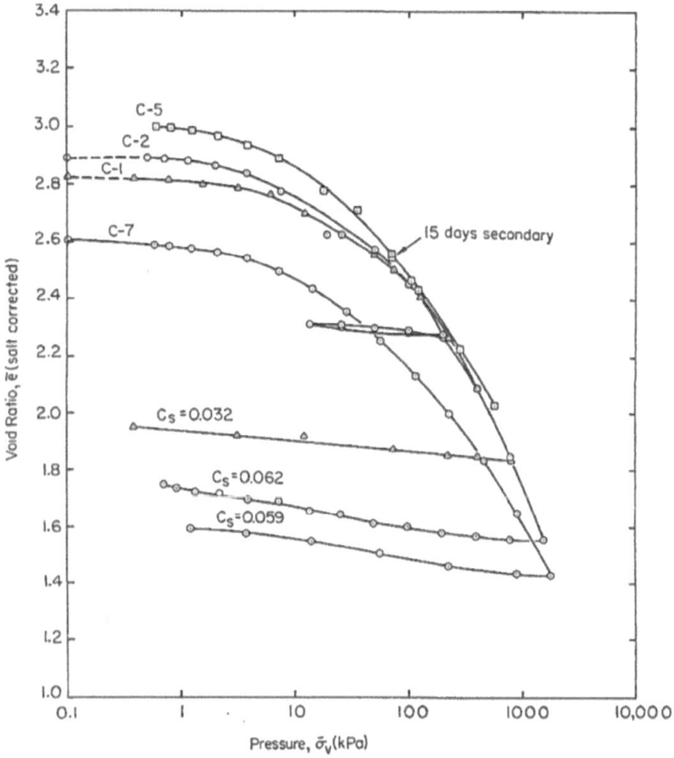

FIGURE 6.30 Consolidation test result of a Caribbean calcareous ooze, load increment ratio $\Delta\sigma_v' / \sigma_v'$ of 1 (Valent et al., 1982). Reprinted with permission of American Society for Testing and Materials.

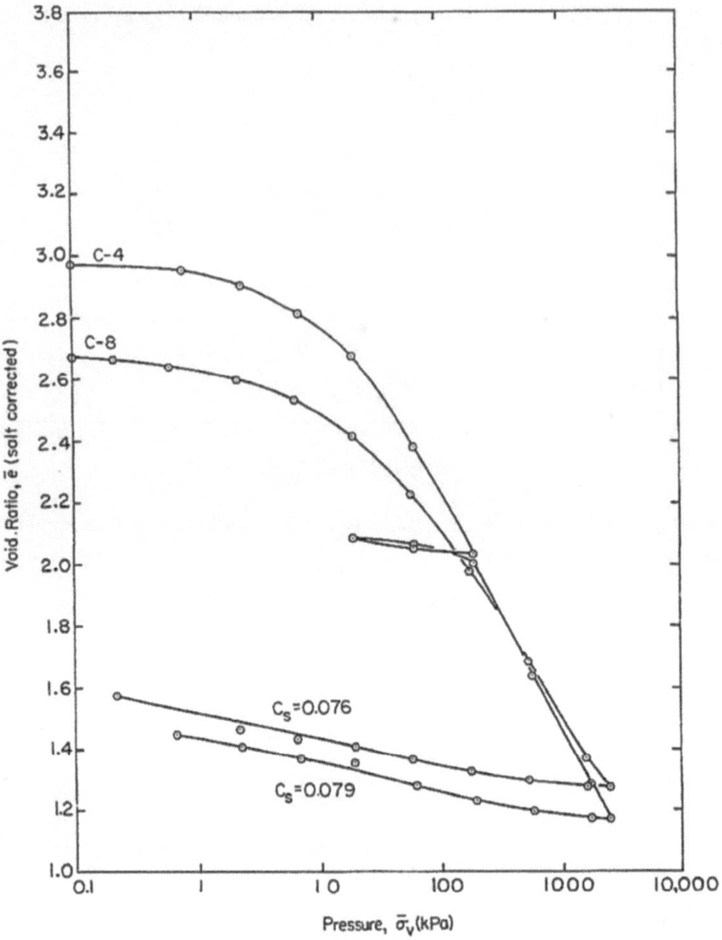

FIGURE 6.31 Consolidation test result of a Caribbean calcareous ooze, load increment ratio $\Delta\sigma_v' / \sigma_v'$ of 2 (Valent et al., 1982). Reprinted with permission of American Society for Testing and Materials.

$\Delta\sigma_v' / \sigma_v' = 2$, Figure 6.30. This suggests that the e versus log σ_v' curve shape differences are possibly the result of test procedure.

In general the following results can be made on compression behavior of deep sea marine calcareous sediments.

- High-carbonate sediments do not compress to a low final void ratio as do low carbonate sediments
- A trend toward granular compression behavior is observed with increasing carbonate content.
- Calcareous soils exhibit slightly over-consolidated behavior possibly a result of cementation and aging.

- There is a good empirical correlation between the compression index (C_c) and plasticity index ($C = 0.024 + 0.014\text{PI}$) since the plasticity index nullifies the effect of intra-particle water (Demars, 1982).

6.3.4.2 Secondary Consolidation

Secondary consolidation occurs when the soil structure is susceptible to creep deformation under a constant stress. The deformation results from the fabric elements adjusting slowly to a more stable equilibrium. This mechanism involves sliding at interparticle contacts, and expulsion of water from micro pore spaces. In nature, chemical, biological, and climatological changes also develop over long time periods. These changes can accelerate the establishment of equilibrium or create new conditions of disequilibrium. The rate at which this occurs is controlled by how fast the rate at which the structure can deform.

During secondary consolidation, the relationship between the void ratio and the log of time is usually a linear line sloping downward over the range of times of interest. There is no reason to believe that secondary consolidation can continue indefinitely. The final equilibrium of soil structure will develop under a given stress state. This relationship is described by the secondary compression index ($C\alpha$). This relationship has been defined by Raymond and Wahls (1976) and Mesri and Godlewski (1977) as given by the following equation.

$$C_\alpha = \frac{de}{d(\log t)} \tag{6.20}$$

In addition, a modified secondary compression index ($C_{\alpha e}$) was defined by Holtz and Kovacs (1981) and is defined as follows:

$$C_{\alpha e} = \frac{C_\alpha}{1 + e_p} \tag{6.21}$$

where

C_α is the secondary compression index

e_p is the void ratio at the start of secondary consolidation

Various authors have shown that the ratio $\frac{C_{\alpha e}}{C_c}$ is approximately constant for NC clays over the normal range of engineering stresses. Typical $\frac{C_{\alpha e}}{C_c}$ for a variety of soils is presented in Table 6.6. Average values of $\frac{C_{\alpha e}}{C_c}$ are as follows (Mitchell, 1976) and Poulos (1988):

- Bass Strait calcareous sands 0.01–0.03
- Inorganic clays and silts: 0.04 ± 0.01
- Organic clays and silts: $0.05 + 0.01$
- Peats: $0.075 + 0.01$.

Mesri and Godlewski (1977) provided a table of values of the ratio of coefficients of secondary compression to compression index for natural soils, Table 6.6.

TABLE 6.6
Values of the Ratio of Coefficient of Secondary Compression to Compression Index for Natural Soils

Grouping	Soil Type	C_α/C_c
Inorganic clays and silt	Whangamarino clay	0.03–0.04
	Leda clay	0.025–D.06
	Soft blue day	0.026
	Portland sensitive clay	0.025–0.055
	San Francisco bay mud	0.04–0.06
	New Liskcard varved day	0.03–0.06
	Silty clay	0.032
	Nearshore clays and silts	0.055–0.075
	Mexico City clay	0.03–0.035
	Hudson River silt	0.03–0.06
Organic clays and silts	Norfolk organic silt	0.05
	Calcareous organic silt	0.035–0.06
	Postglacial organic day	0.05–0.07
	Organic clays and silts	0.04–0.06
	New Haven organic day silt	0.04–0.075
Peats	Amorphous and fibrous peat	0.035–0.083
	Canadian muskeg	0.09–0.10
	Peat	0.075–0.085
	Fibrous peat	0.06–0.085

Source: Mesri and Godlewski (1977). Reprinted with Permission from ASCE.

The range of the ratio $\frac{C_{\alpha e}}{C_c}$ tends to decrease as the initial void ratio increases. $C_{\alpha e}$ increases considerably as the vertical effective stress σ_v' increases, and for one of the Bass Strait soils can be roughly approximately as follows:

$$C_{\alpha e} = 0.00077\left(\log_{10}\sigma_v' - 1\right) \tag{6.22}$$

Mesri (1973) has also provided another method to obtain the secondary compression index for clays, silts, and organic materials, and it is shown in Figure 6.32. The ($C_{\alpha e}$) is plotted versus the natural water content of the soil.

A comparison of the consolidation characteristics of different sediment types is presented in Figure 6.33. A review of Figure 6.33 shows the following: the silt and foram/nanno ooze both exhibit relatively flat consolidation curves, and the clays and nannofossil marl all exhibit a rather steep virgin consolidation curves when compared to the silt/foram/nanno ooze curves. The lack of a knee on a number of the curves could either indicate sample disturbance or initial consolidation.

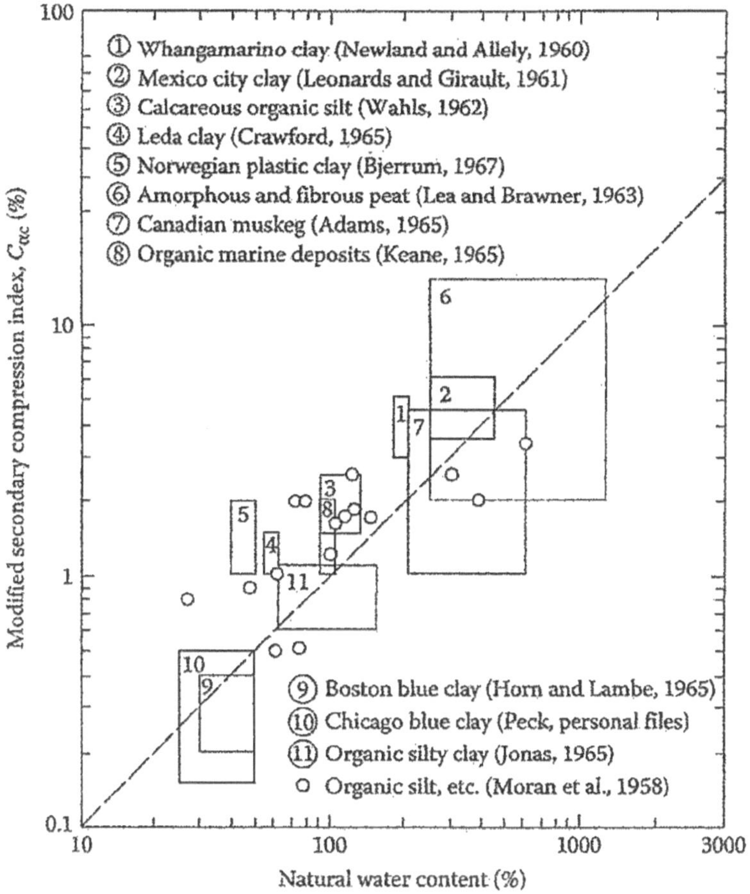

FIGURE 6.32 Secondary compression index for clays, silts, and organic materials (Mesri 1973). Reprinted with permission from ASCE.

6.3.5 Permeability

The void spaces between soil grains allow liquids to flow through them. A measure of this ability of a soil to transmit liquids is its permeability. Factors affecting permeability are (1) the effective grain size or effective pore size, (2) shapes of voids and flow paths through the soil pores—tortuosity, (3) saturation, and (4) viscosity of permeant. The majority of these issues are covered in detail in other books such as Chaney and Almagor (2016) and will not be revisited here. In this section results of permeability tests on calcareous sediments will be discussed. A comparison of void ratio versus permeability for a variety of sediments is presented in Figure 6.34 (Silva et al., 1981; Schultheiss and Gunn, 1985). A review of this figure shows that there are two general trends. The first trend is for increasing impermeability from a silt-like material to foram ooze to clays. The second trend is increasing permeability for all materials as their void ratio increases.

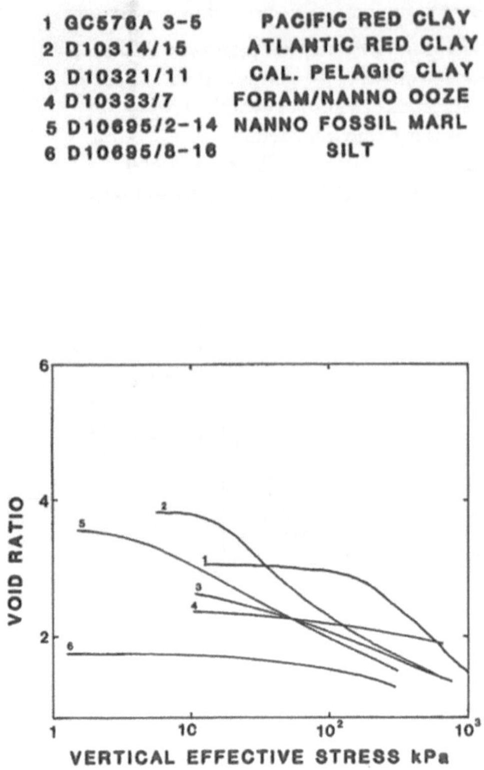

FIGURE 6.33 A comparison of the consolidation characteristics of different sediment types (Schultheiss and Gunn 1985).

6.3.6 Shear Strength

The shear strength of calcareous soils can be modeled in three stages depending upon the amount of calcium carbonate present and whether it is cemented or not. Initially, there would be a phase in which a few carbonate particles would be present in a soil matrix. The soil matrix at this stage would dominate the resistance to shearing stresses. The calcium carbonate particles would simply be carried along during the shearing process. As the amount of calcium carbonate increases a framework would be developed to carry the shearing stresses. At this stage, the soil matrix would become irrelevant. Strength would result from the interaction of carbonate particles. The sediment would behave similarly to a granular material. In the intervening transition both the matrix and carbonate particles would contribute to the strength. A final stage would occur when the individual carbonate particles become cemented together.

Lee (1982) has presented results from a test program on eastern equatorial pacific sediments with less than 70% calcium carbonate. The 70% was obtained by

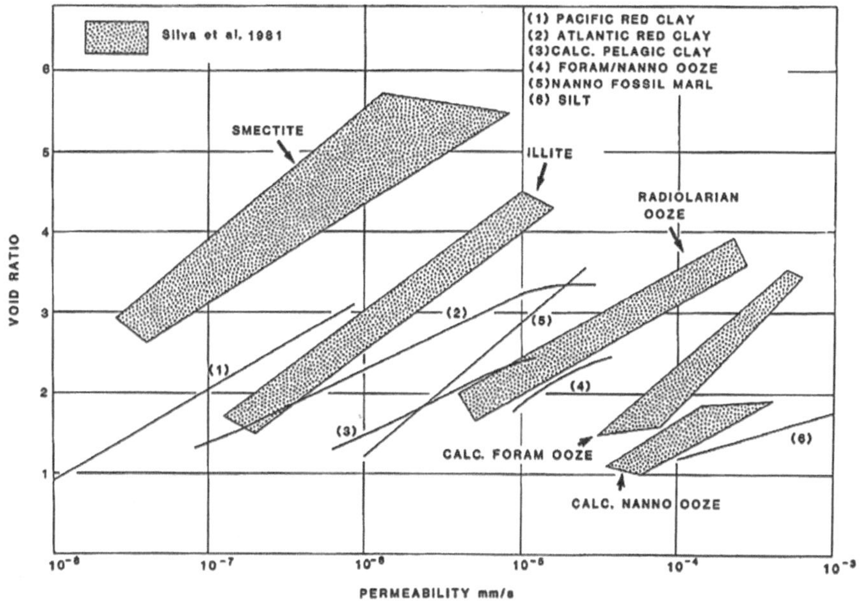

FIGURE 6.34 A plot of void ratio versus permeability (Silva et al., 1981; Schultheiss and Gunn, 1985).

Lee (1982) based on density tests discussed earlier. This percentage is well below the zone of matrix-carbonate interaction.

Mechanical versus chemical sources of strength is a complex topic in calcareous sediment. As an example, a graph of vane shear strength versus overburden stress for a sample from the eastern equatorial Pacific is presented in Figure 6.35. A review of this figure implies that the in-situ stress state determines shear strength as it should if the material behavior were essentially frictional. But in this case the overburden stress has been shown to increase approximately with age. Therefore an older sediment may be expected to be more cemented than a younger sediment. Given this information the source of strength cannot be determined.

Poulos et al. (1982) conducted drained triaxial compression (CID) tests on remolded carbonate sand. The sand was of biogenic origin from Bass Strait, Australia. The axial stress-strain and volume strain versus axial strain curves are presented in Figure 6.36. A review indicates that the behavior is similar to silica sand.

When stresses are increased to a magnitude where significant soil grain crushing occurs, the effective angle of internal friction will decrease compared to the angle determined when little grain crushing is taking place. Figure 6.37 is a plot of shear stress (½ principal stress difference) versus normal stress (½ principal stress sum). A review of this figure shows that as the normal stress increases the angle of the failure envelope decreases. This is caused by soil grain crushing.

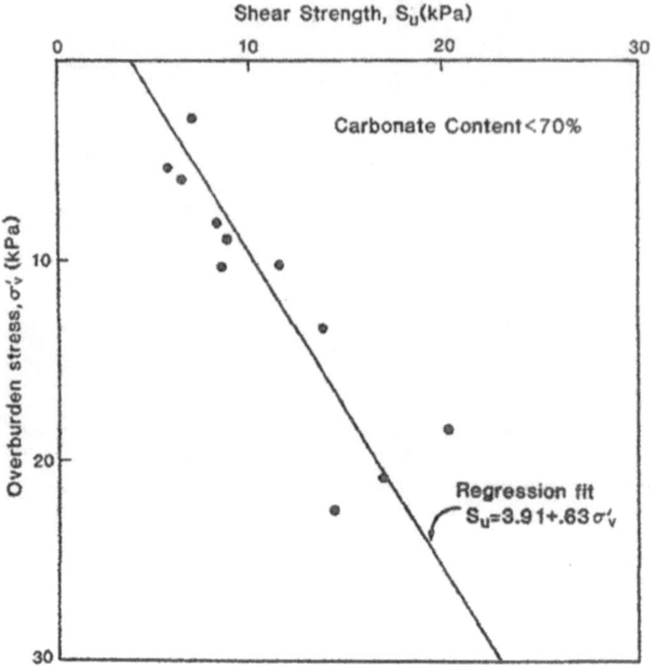

FIGURE 6.35 Correlation of laboratory vane shear strength with in situ overburden stress, eastern equatorial Pacific specimens with carbonate contents less than 70% (Lee, 1982). Reprinted with permission of American Society for Testing and Materials.

In Figure 6.38 stress paths for undrained triaxial shear tests, on remolded specimens of foram shells, indicate the development of high excess pore pressures during shearing. This is indicated by the sharp decrease in normal stress (i.e., principal stress sum) at near constant shear stress (i.e., principal stress difference) (i.e., the long near horizontal portion of the stress strain curve). The specimen during this phase is undergoing a form of structural rearrangement (i.e., reorganization). During this phase effective stresses between sediment grains are decreasing and shifting the total stresses to the pore water phase. This behavior is either due to the rearrangement of sediment grains to decrease the number and effectiveness of grain contacts or the sediment grains themselves (foram shells) are puncturing and crushing.

To evaluate the effect of foram ooze on piles Valent ran residual direct shear tests using calcareous sands and silts. These tests consisted of three reversed loading large displacement runs. The results are presented in Figure 6.39. A review of Figure 6.39 demonstrates that the angle of friction between soil and pile does not degrade for foram ooze. This implies that should low skin friction be observed in pile load tests, it must be due to a very low in situ horizontal stress.

Demars et al. (1976) also investigated the effect of various carbonate materials on effective stress shear strength parameters as shown in Table 6.7. A review shows that he reported effective angles of friction varying from 27.7° to 51.0° corresponding to cohesion ranging from 0 to 9 kPa.

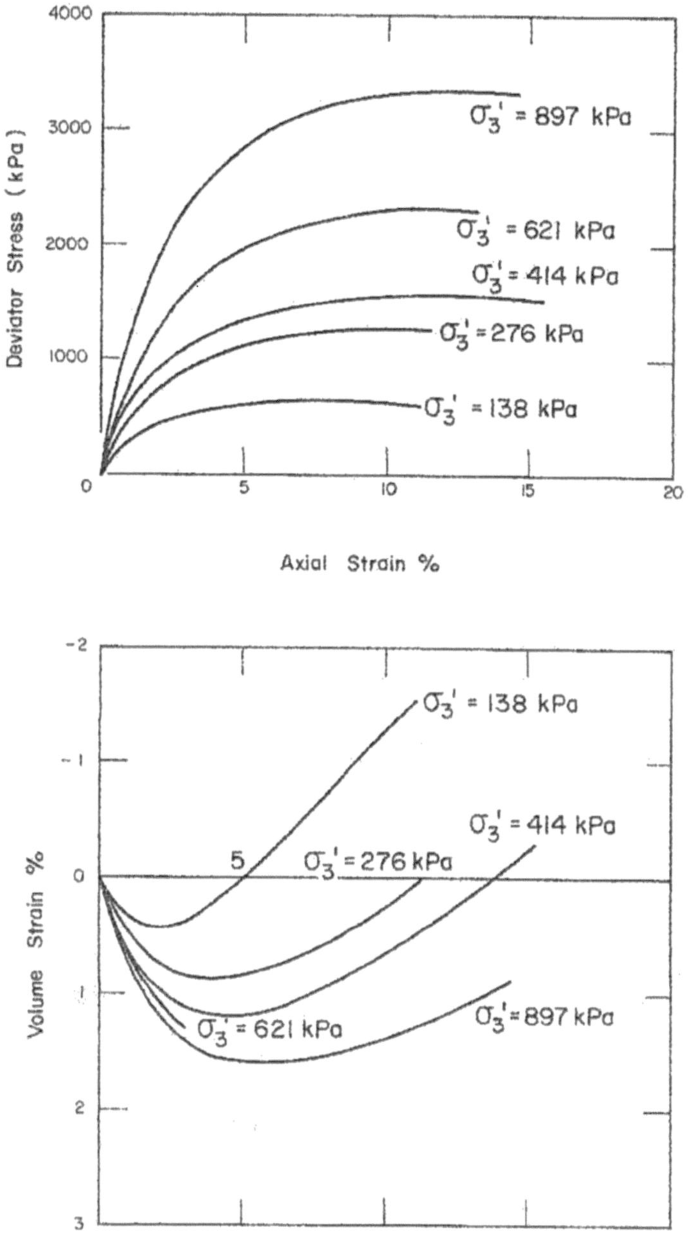

FIGURE 6.36 Remolded drained triaxial compression tests on carbonate sand of biogenic origin from Bass Strait, Australia (After Poulos et al., 1982).

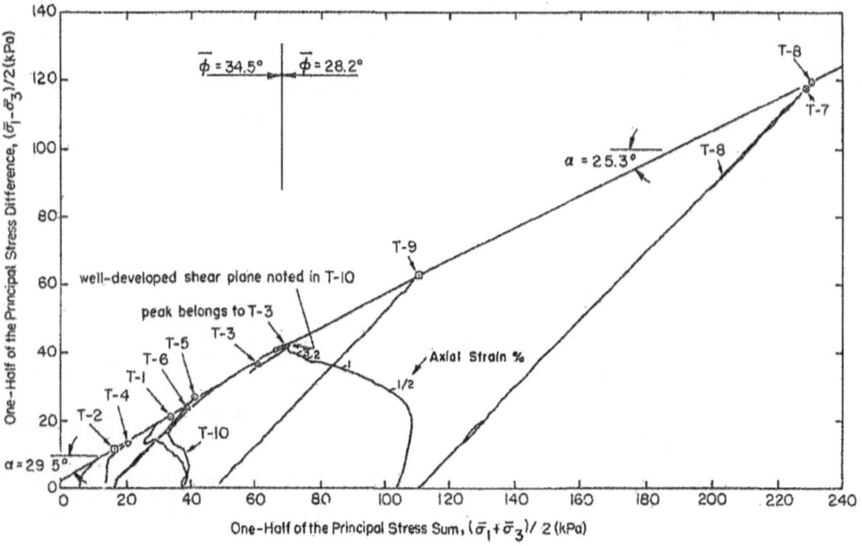

FIGURE 6.37 Stress path curves and failure envelop for CIU and CID triaxial specimens, Caribbean (Valent et al., 1982). Reprinted with permission of American Society for Testing and Materials.

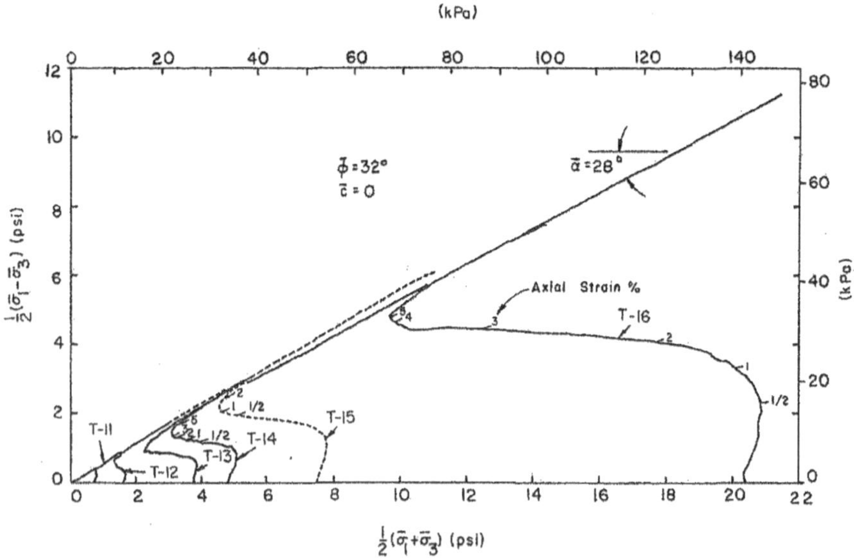

FIGURE 6.38 Stress path curves and failure envelope for CIU tests of Blake Plateau foram ooze (Valent et al., 1982). Reprinted with permission of American Society for Testing and Materials.

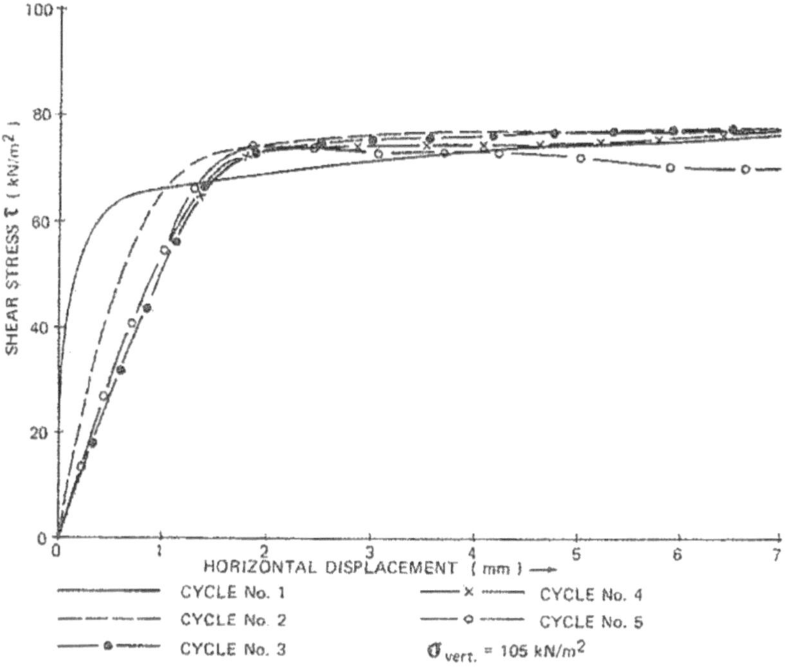

FIGURE 6.39 Repeated drained direct shear tests on calcareous silt (North Rankin) and steel (roughness μ = 4.5–5.0) (Beringen et al. 1982). Reprinted with permission of Taylor & Francis.

6.3.7 Deformation Parameters

Poulos et al. (1982, 1984) have presented the secant Young's modulus E_s and Poisson ratio υ_s results from tests on carbonate material as shown in Figures 6.40 and 6.42. These test results are from drained triaxial tests consolidated either under hydrostatic or K_o consolidated conditions. A review of Figure 6.40 shows that the variation of E_s with mean initial effective stress for three applied stress levels increases linearly with increasing $p_o^{'}$, and decreases as the applied stress level increases. In Figure 6.41 deformation parameters from remolded drained triaxial compression tests carbonate sand of biogenic origin from Bass Strait, Australia. Both $E_{50}^{'}$ and $v_{50}^{'}$ were found to vary linearly with the effective confining pressure $\sigma_3^{'}$. The relationships may be expressed as follows:

$$E_{50}^{'} = 5.5 + 75\sigma_3^{'} \tag{6.23}$$

$$\nu_{50}^{'} = 0.29 - 0.57\sigma_{50}^{'} \tag{6.24}$$

where $E_{50}^{'}$ and $\nu_{50}^{'}$ are in MPa.

TABLE 6.7
Effective Stress Strength Parameters of Calcareous Soils

Reference	Soil Type	Carbonate Content ('¥o	Location	Test Type	Confining Pressure or Vertical Stress (kPa)	c' (kPa)	ф>′ (degrees)	Remarks
Nacci et al. (1974)	Siliceous calcilutite	20.65	Labrador Basin	CIU		2–7	31–37	Cemented samples; confining pressures slightly greater than effective overburden pressure
Datta et al. (1979)	Bioclastic carbonate sand (four types)	:>85	Three from west coast of India; one from islands in Arabian Sea	CID	100	0	49.5–51.0	Similar values for all four soils; c', ϕ depend on confining stress and crushing; c', ф' values for peak stress ratio
				CID	15000	3–9	29–30	
				CID	100	0	42–44.5	
				CID	6400	a	40.5—42	
Poulos et al. (1982)	Carbonate sand	88	Bass Strait, Australia	CID	138–897	0	46.3–40.4	ϕ ' decreased with increasing confining pressure
Demars et al. (1976)	**Various**	**Various**	**Various**	CIU	7–70	0–1	27.7–31.3	Effect of carbonate content examined

Source: Demars et al. (1982).

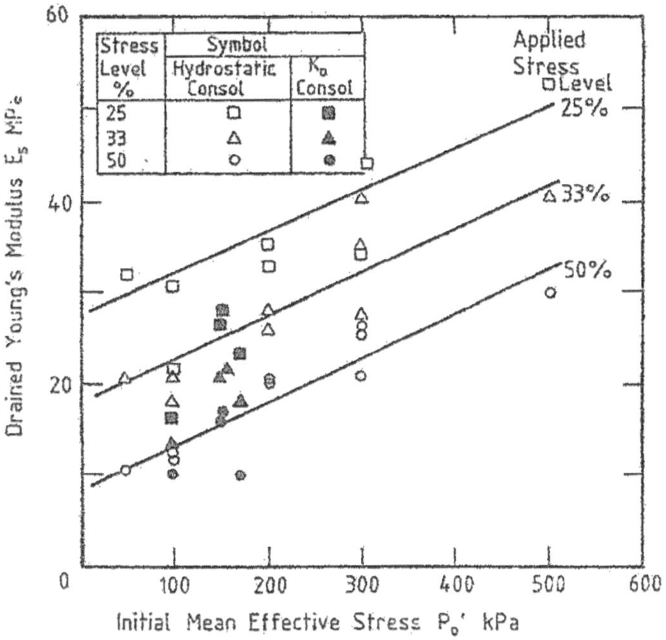

FIGURE 6.40 The drained Young's modulus for Bass Strait carbonate sand (Poulos et al., 1984).

In contrast, the variation of the drained Poisson ratio υ_s with mean effective stress p_o' on the same material as above is shown in Figure 6.42. A review shows that the υ_s decreases with increasing p_o' but is almost independent of the applied stress level.

6.4 SILICEOUS SEDIMENTS

6.4.1 INTRODUCTION

Siliceous biogenic sediments are composed of opal silica (hydrated silica dioxide: $SiO_2 \bullet nHO_2$). Silica (S_i) is found in the skeletons of benthic organisms (sponges) on the continental shelves, and planktonic organisms on the slope and in deep water. Siliceous oozes cover approximately 15% of the deep sea floor. Oozes are defined as sediments that contain at a minimum 30% skeletal remains of pelagic microorganisms. Silica (Si) is a bio-essential element and is efficiently recycled in the marine environment through the silica cycle.

Siliceous sediment materials in the deep sea are typically frustules (hard siliceous shells). These frustules are of diatoms (Figure 6.43), phytoplankton siliceous algae (ranging in size from a few micrometers to about 2 mm) and radiolarian (Figure 6.44), a group of tiny protozoans. Radiolarians have various types of complex skeletons with typical sizes ranging from a few tens to a few hundreds of micrometers (Figure 6.44). The particle sizes of typical siliceous ooze microfossils fall in the silt size range of 0.03–0.07 mm.

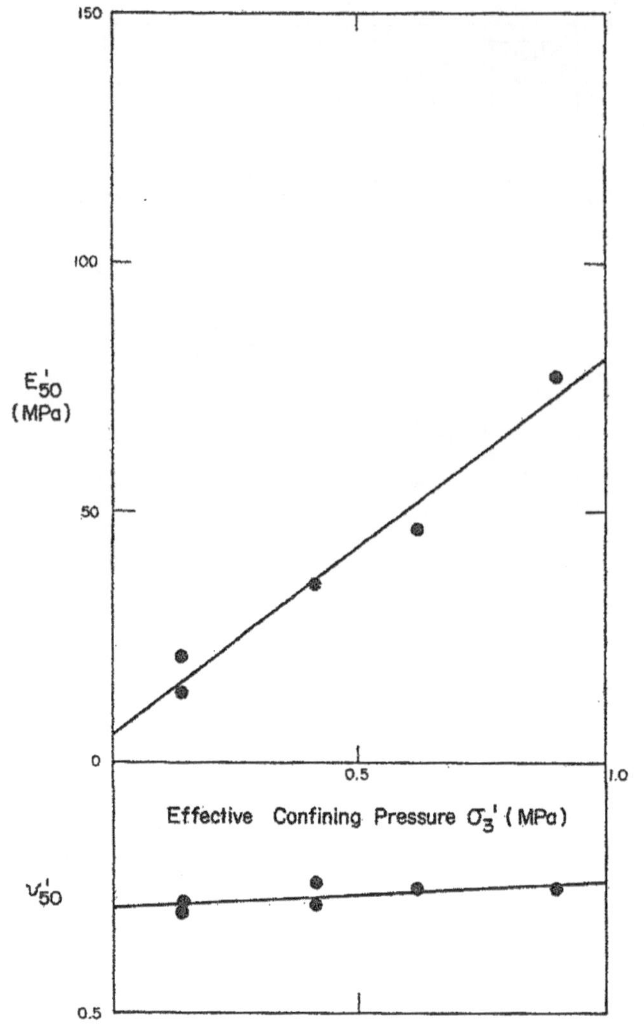

FIGURE 6.41 Deformation parameters from remolded drained triaxial compression tests carbonate sand of biogenic origin from Bass Strait, Australia (Poulos et al., 1982).

Diatom oozes are common in high latitudes. Diatom muds (mixture of diatoms and terrigenous muds composed of terrestrially derived silica particles and sponge spicules) occur near continental margins. In contrast, radiolarian ooze is common in the equatorial regions. A number of factors affect the opal silica content in seawater and the occurrence of siliceous oozes. These factors (1) distance from land masses, (2) water depth, and (3) ocean fertility. Similar to calcareous skeletons, siliceous microfossils are also subject to dissolution in deeper waters. The actual factors controlling the dissolution are complex; but in general, diatoms dissolve more easily than radiolarians.

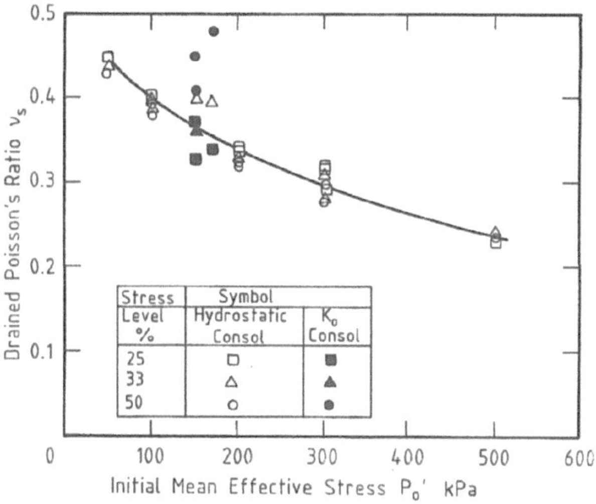

FIGURE 6.42 The drained Poisson ratio for Bass Strait carbonate sand (Poulos et al., 1984).

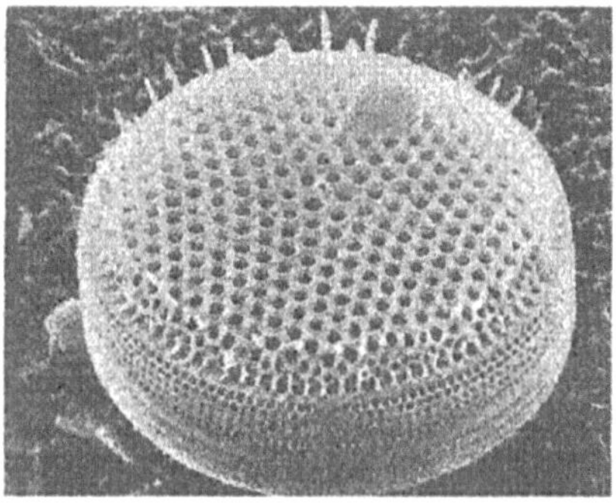

Magnification Factor: 4300

FIGURE 6.43 Siliceous diatom (Noorany, 1983) Note: Most of the skeletons are approximately ¼ to ⅓ mm in size.

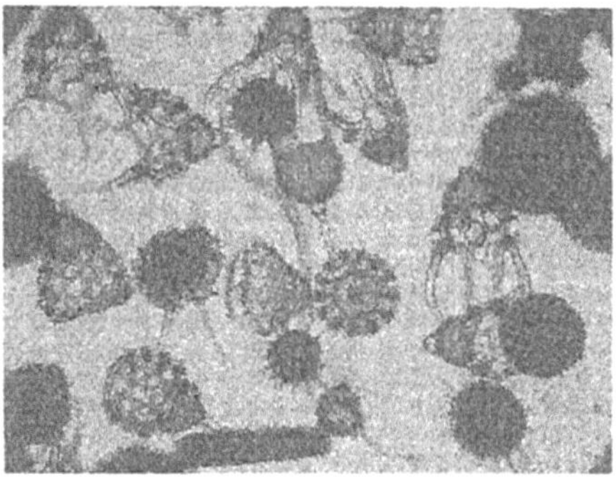

FIGURE 6.44 Assemblage of radiolarians (Noorany, 1983).

Silicoflagellate is another siliceous material which is often found as a secondary component of radiolarian and diatom oozes. Silicoflagellates are a planktonic marine organism with skeletons made of hollow rodlets that are between 20 to 50 μm in diameter.

6.4.2 Density and Water Content

Density and water content data of siliceous diatom oozes from the Bering Sea and the Sea of Japan have been reported by Davie et al. (1978). High water contents were found, ranging between 89 and 205%, with an average value of 135%. No trend of decreasing water content with increasing penetration below the seabed was observed. Hamilton (1976) found that, even at a penetration of 500 m, the water content of a Bering Sea diatom ooze was still about 105%.

Average values of specific gravity for siliceous oozes are about 2.45. However, values as a low as 2.3 have been reported, since they consist primarily of opaline silica with a specific gravity of 2.10 (Poulos, 1988).

Horn et al. (1976) has reported the following average data for radiolarian ooze from the North Pacific Ocean:

- Average saturated unit weight 11.5 kN/m^3
- Water content 340%
- Porosity 88%.

In contrast, pure radiolarian oozes had the

- Lowest unit weight (average 11.2 kN/m^3)
- Highest porosity (average 89%)
- Highest moisture content (average 389%) of any ocean sediment.

Microscopic examination of this soil revealed not only high interstitial porosity, but also grains which were porous and hollow (Table 6.8).

6.4.3 Consistency

A sediment with increasing opal silica (hydrated silica dioxide: $SiO_2 \bullet nHO_2$) content exhibits decreasing plasticity and behaves like a frictional material.

6.4.4 Compressibility

High values of the compression index C_1 are evident from the data collected by Davie et al. (1978), which is plotted in Figure 6.43. As with other soil types, there is a clear trend of increasing C_1 with increasing initial void ratio e_o.

6.4.5 Permeability

The void spaces between soil grains allow liquids to flow through them. A measure of this ability of a soil to transmit liquids is its permeability. Factors affecting

TABLE 6.8
Typical Range of Siliceous and Radiolarian Oozes

Material Type	Location	Penetration (m)	Water Content Range (%)	Water Content Ave. (%)	Specific Gravity	Unit Weight (kN/m)	Porosity (%)	Ref.
Siliceous Diatom Ooze	Bering Sea and Sea of Japan		89–205	135				Davie et al. (1978)
Diatom Ooze	Bering Sea	500		105				Hamilton (1976)
Siliceous Ooze					2.3–2.45			Poulos (1988)
Opaline Silica					2.1			Poulos (1988)
Radiolarian Ooze				340		11.5	88	Horn et al. (1976)
Pure Radiolarian Ooze				389		11.2	89	Poulos (1988)

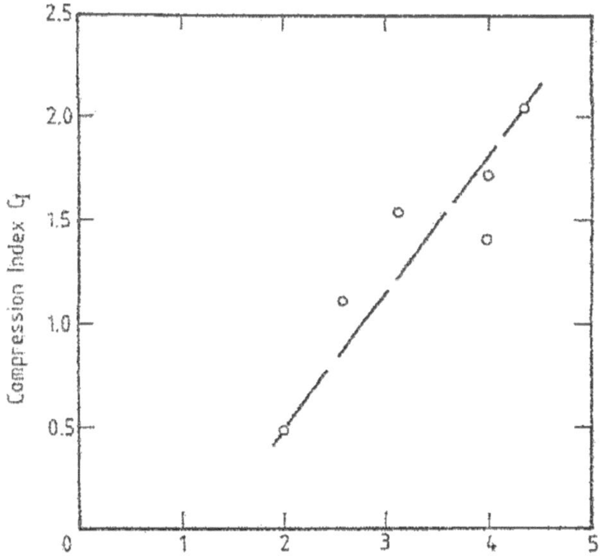

FIGURE 6.45 Compressibility of diatom ooze (after Davie et al. 1978). Reprinted with permission of Taylor & Francis.

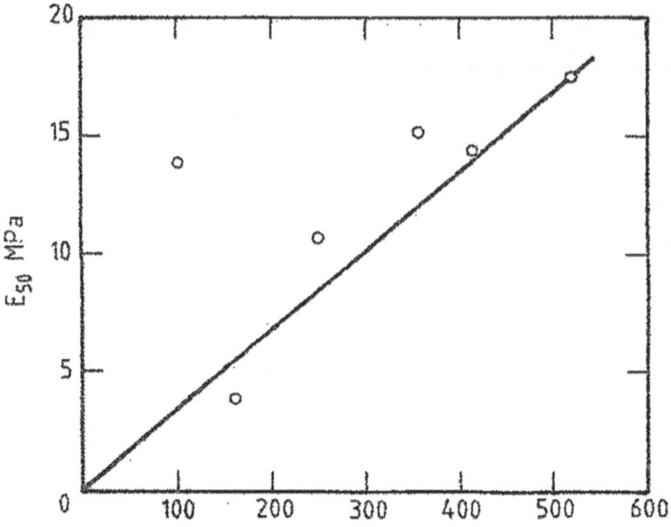

FIGURE 6.46 Undrained modulus for diatom ooze (after Davie et al., 1978). Reprinted with permission of Taylor & Francis.

permeability are (1) the effective grain size or effective pore size, (2) shapes of voids and flow paths through the soil pores—tortuosity, (3) saturation, and (4) viscosity of permeant.

6.4.6 Shear Strength

Davie et al. (1978) reported very low diatom ooze strengths may arise from the absence of clay binder in the ooze and small submerged unit weight (typically half that of calcareous oozes). Strengths as low as about 35 kPa at 150 m penetration (increasing linearly with depth) have been reported. The CU triax test with pore pressure measurement indicates values of effective stress friction angle ϕ of 36° and 41° for two samples of diatom ooze. A graph of Young's modulus as a function of the initial mean effective stress is presented in Figure 6.46. A review of this figure shows that the Young's modulus (i.e., stiffness) of diatom ooze increases linearly with mean effective stress.

6.5 ORGANIC SEDIMENTS

6.5.1 Introduction

Organic substances in quaternary sediments consist of macroscopic matter (i.e., fruits, leaves, roots, tree parts, and animals) and microscopic matter (i.e., fragments of plants and animals, pollen, micro-organisms, organic molecules, and compounds). In addition, dead, dormant, and living organisms can co-exist in the sediments. The typical characteristics of organic soils are the following: (1) have low shear strength, (2) tend to be excessively compressible, and (3) tend to degrade with time. Organic matter is a common constituent of marine sediments; it frequently constitutes up to 2% or more by the weight of sediments, especially on continental margins. It exists as biota, including living organisms such as microbes and macroscopic infauna, and as detritus derived from an innumerable variety of dead organisms. It exists in a variety of gross forms including filaments, mats, and particles, and in a variety of chemical forms, including proteins, carbohydrates, lipids, and lignins. Organic matter is thus a general term and is used herein to describe all organic and organic-related components. A summary of much of the information on organic matter has been presented by Hunt (1979). The effect of different types and amounts of organic matter on soil properties has been the subject of a number of investigations. A thorough study of some sensitive Swedish clays and other Swedish soils by Pusch (1973) has detailed the complexity of the possible organic matter–clay relationships and demonstrated the profound impact that organic matter can have on the index properties, deformational characteristics, and strength parameters of a soil.

Humus is the dark organic matter that forms in the soil when plant and animal matter decays. Humic acid produced by biodegradation of dead organic matter or organic molecules such as cellulose in the presence of polyvalent cations form gel complexes that appear to have a bonding effect. In clay and clay/silt sediments macroscopic organic matter is typically of minor importance for geotechnical properties. In contrast, this type of material may be of major importance in peat and mud. Organic matter absorbs water and causes clay-size particles to aggregate forming an open fabric. An example of a classification system for organic material is presented in Table 6.9.

6.5.1.1 Gel

A **gel** has a consistency of a solid jelly-like material that can have properties ranging from soft and weak to hard and tough. Gels are defined as a dilute cross-linked system. When in a steady-state this system has no flow characteristics. By weight, gels are mostly liquid, yet they behave like solids due to a three-dimensional cross-linked network within the liquid. It is the crosslinking within the fluid that gives a gel its structure (hardness) and contributes to the adhesive stickiness. In this way gels are a dispersion of molecules of a liquid within a solid in which the solid is the continuous phase and the liquid is the discontinuous phase.

6.5.1.2 Marine Clays and Organic Material in South East Asia

The organic content of sediments from around the South China Sea is presented in Figure 6.47 (Cox, 1970). A review of Figure 6.47 shows that the range of percent organics in the Gulf of Thailand ranges from 2 to 4. In contrast, the range of organics from highway projects in the coastal zone of Thailand ranges from approximately 2.3% to 5%. These organic contents are higher than those normally found for recent post glacial marine clays in the temperate climates of the northern hemisphere.

TABLE 6.9
Classification Systems for Organic Soils

Major Divisions	Group Symbol	Organic Content (%)	pH Range	Specific Gravity	Drainage Char.	w_L (%)	w_p (%)	Value as Base Mat.	Value as Embankment Mat.
Sandy clay and organic mat.	SWO-1E	0–15	>7.0	>2.4	Pervious to fair	26–37	7–8	Excellent to fair	Excellent
	SWO-2F	15–20	6.0–7.0	2.18–2.40	Poor	37–48	3–8	Fair to poor	Fair to poor
	OS-1F	21–30	6.0–7.0	2.18–2.40	Poor	37–48	3–8	Fair to poor	Fair to poor
	OS-2	31–50	5.5–6.0	1.83–2.18	Poor	44–72	0.8–3	None	Poor
	OS-3	>50	<5.5	<1.83	Poor	>72	—	None	None
Silt and organic mat.	SWO-1G	0–5	>7.0	>2.45	Impervious to very poor	32–41	12–13	Good to fair	Excellent to Fair
	SWO-2F	15–20	6.0–7.0	2.10–2.45	Very poor to poor	42–56	13–20	Fair	Fair
	OS-1F	21–30	6.0–7.0	2.10–2.45	Very Poor to poor	41–56	13–20	Fair	Fair
	OS-2F	31–50	5.5–6.0	1.75–2.10	Poor	56–72	17–20	Fair to poor	Poor
	OS-3	>50	<5.5	<1.75	Poor	>72	—	None	None
Clay and organic mat.	SWO-1	0–15	>7.0	>2.37	Impervious to very poor	65–67	26–40	None	Fair to poor
	SWO-2	15–20	6.0–7.0	2.15–2.37	Very poor to poor	65–66	18–6	None	None
	OS-1	21–30	6.0–7.0	2.25–2.37	Very poor to poor	65–66	18–26	None	None
	OS-2	31–50	5.5–6.0	1.82–2.15	Poor	66–76	22–13	None	None
	OS-3	>50	<5.5	<1.82	Poor	>76	—	None	None

Source: Modified from Arman (1970).

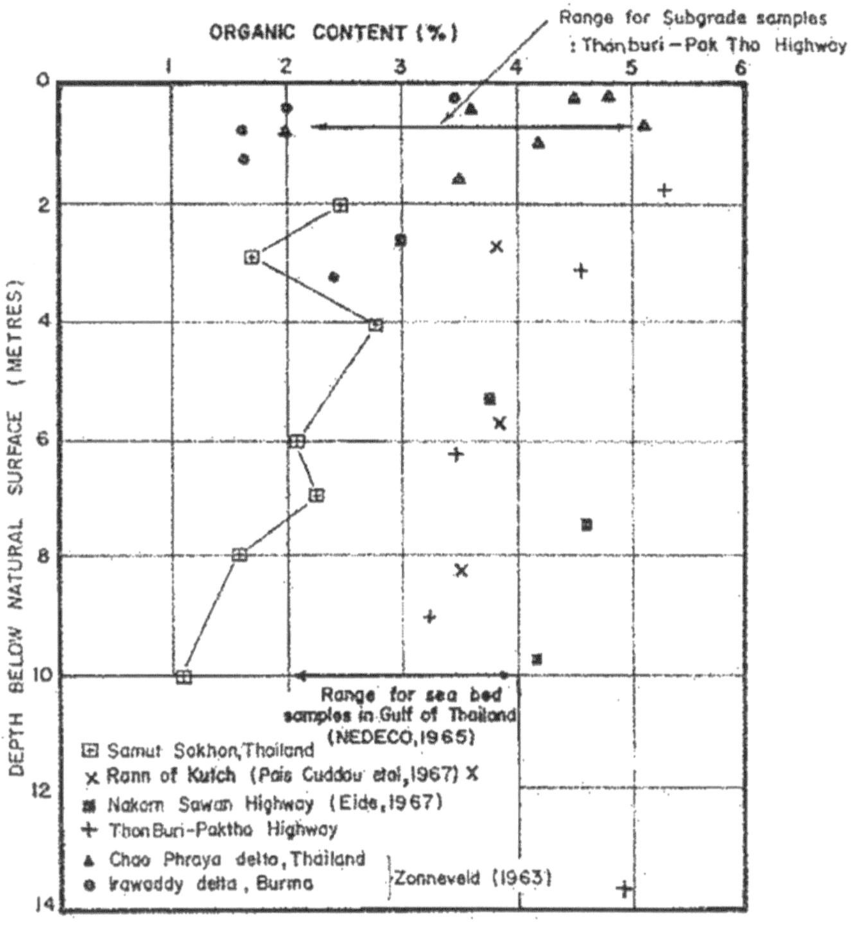

FIGURE 6.47 Organic content determinations for recent marine clays in SEA (Cox, 1970).

6.5.2 Density and Water Content

The organic soil can hold high water content as shown in Figure 6.49. It can be seen that a soil with up to 95% organic material can hold as much as 600%–1500% of water. Water content as a function of percent organic content is presented in Figures 6.48 and 6.49.

The aqueous phase of most organic soil is acid with pH values ranging from 4 to 7, but values as low as 2 and as high as 8 have been encountered.

Water content of peaty soil and marine clay as a function of dry density is shown in Figure 6.50. Review of the figure shows that as water content increases from approximately 20% to 1500%, the dry density decreases from 15 to 0.08 gm/cm^3. This is true for both peat and marine clay. The behavior of wood is approximately the

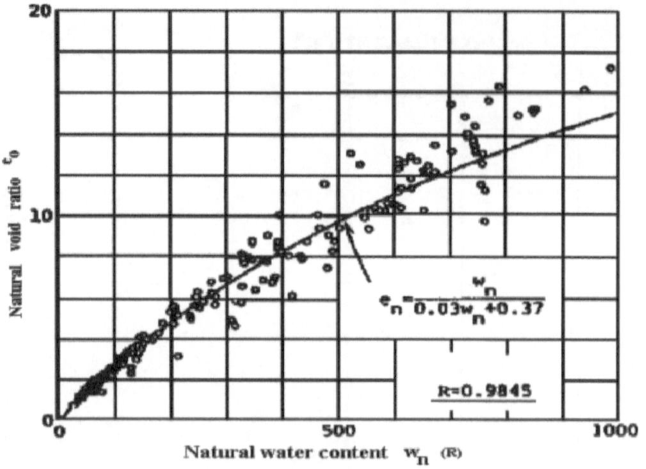

FIGURE 6.48 Relation between natural water content and natural void ratio of peat (Oikawa and Igarashi, 1997).

same after a water content of 100%. Before a water content of 100% the dry density of wood is slightly lower than that of peat and marine clay.

6.5.3 Consistency

The plasticity characteristics of a sediment is characterized by its consistency. Atterberg limits and related indexes are a function of many parameters: drying (Casagrande, 1932), temperature, molding and grain size (White and Walton, 1937), surface area (Farrar and Coleman, 1967), and clay mineral composition (Seed et al., 1964). The main factors are those tied to the physicochemical properties of the clay–water system (Moum and Rosenquvist, 1961; Soderblom, 1969). Generalized plasticity characteristics of sediments from a variety of locations (i.e., prodeltas, deltas, gulfs and bays, continental margins and trenches, etc.) are presented in Figure 6.51 (Chassefiere and Monaco, 1983). A review indicates that both the liquid limit (IL) and the plasticity index (PI) increase with increasing smectite content. This behavior is centered around the A-line on the Casagrande liquid limit chart. An increase in the percentage of organic matter results in the decrease of PI to the approximate upper limit exhibited by peats.

The effect of organic carbon on Atterberg limits and grain specific gravity has been presented by Booth and Dahl (1986) (Figure 6.52). A review of Figure 6.52 shows that as organic carbon increases the Atterberg limits increase and the grain specific gravity decreases. Plots of regression lines on the relationship between organic matter and geotechnical properties of actual marine sediments from a variety of locations are presented in Figure 6.53. A review of Figure 6.53 shows that natural sediments with varying amounts of percent organic content behave in a manner similar to what was shown in Figure 6.52. Pusch noted from his analytical work that progressive failure and deformation were observable at an organic content of 2%–3%

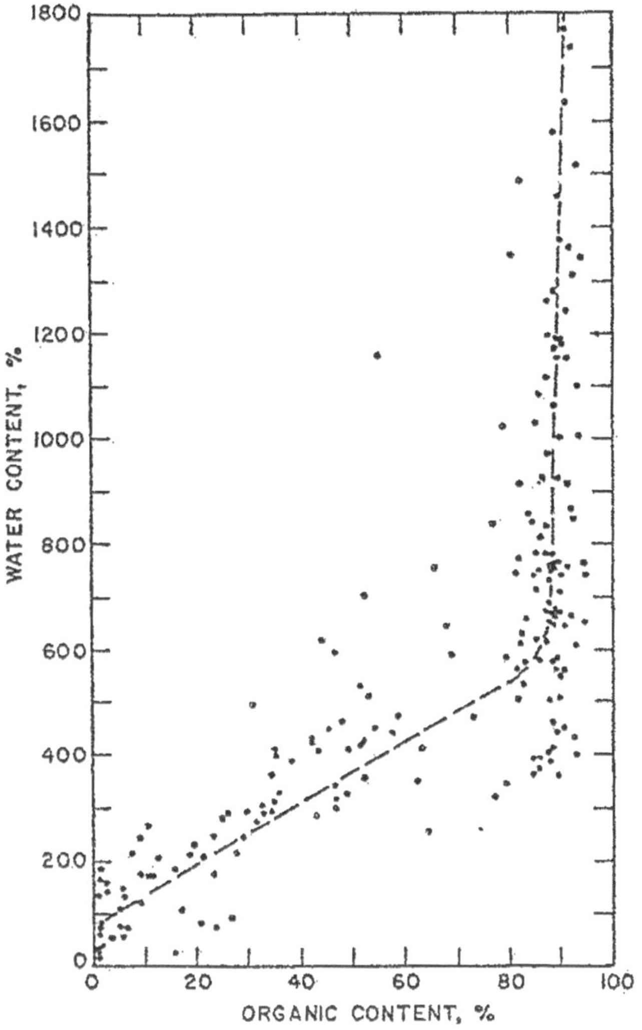

FIGURE 6.49 Organic content versus water content (after MacFarlane and Rutka, 1962). This material is the property of the US government but is in the public domain and is not subject to copyright protection.

and that shear strength was significantly affected at an organic content of 3%–4%. In addition, Pusch showed a dramatic increase in both the liquid and the plastic limit with increasing organic content. This phenomenon was also observed by Odell et al. (1960) using samples from several soil groups and from different soil horizons in Illinois. Similar effects and relationships have been noted for marine sediments. In a study of organic-rich sediments (up to 20% organic carbon) on the Peru–Chile continental slope, Busch and Keller (1981) showed an increase in both liquid and plastic limits with increasing organic content. Reimers (1982) analyzed these same

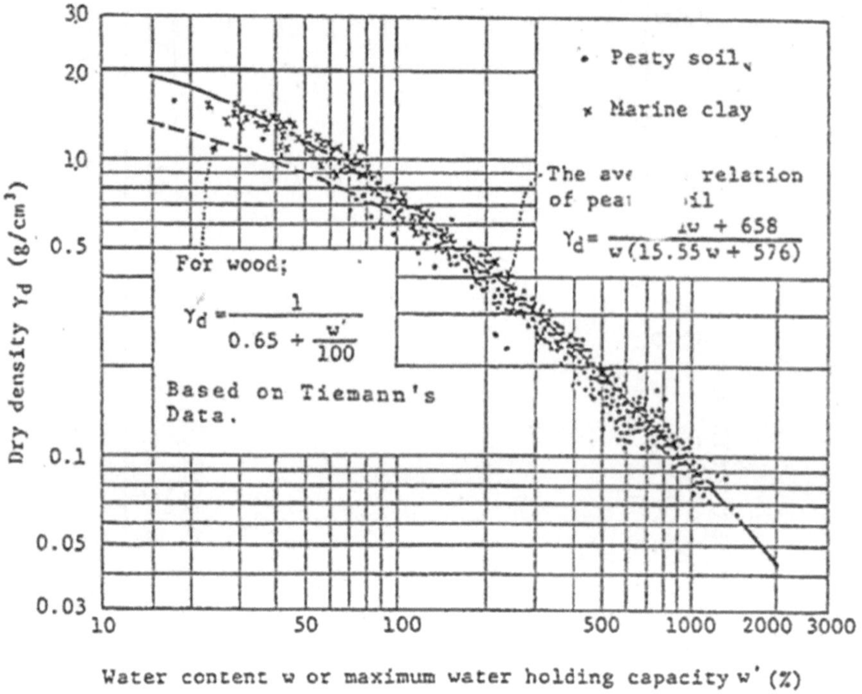

FIGURE 6.50 Water content versus dry density for peaty soil, marine clay, and wood (g/cm^3) (After Ohira, 1977).

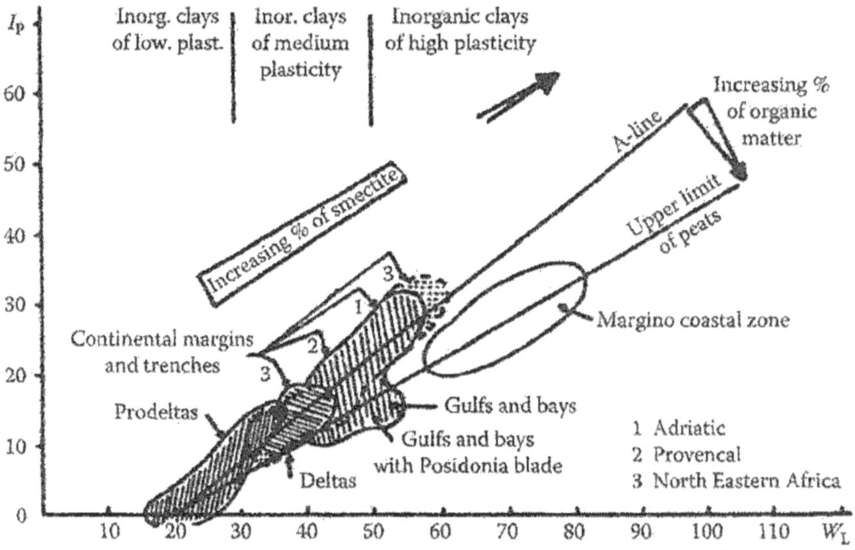

FIGURE 6.51 Plasticity diagram, using seawater, for various marine zones (Chassefiere and Monaco, 1983). Reprinted with permission of Taylor & Francis Group.

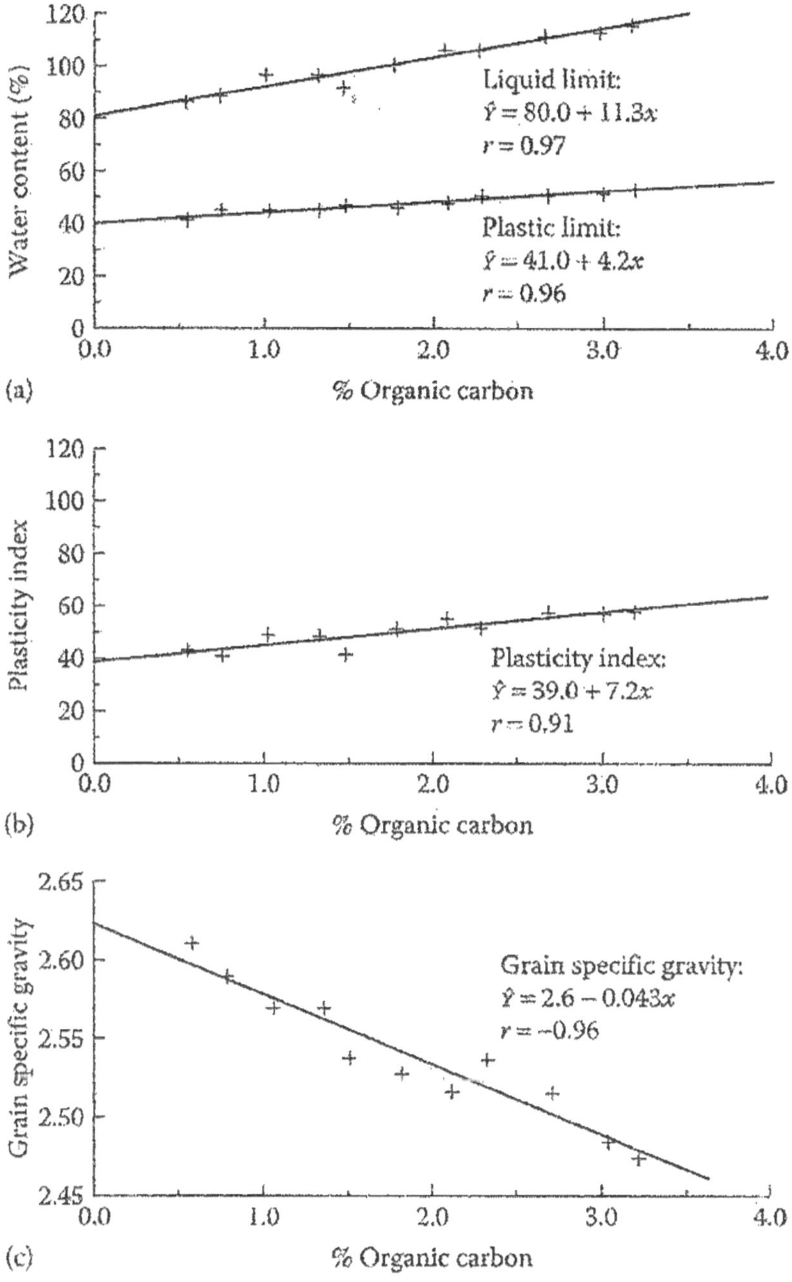

FIGURE 6.52 (a) Liquid limit and plastic limit, (b) plasticity index, and (c) grain specific gravity with increase in organic carbon. Reprinted with permission of Taylor & Francis Group.

sediments and the different types of organic matter that were present. She suggested that changes from one type of organic matter to another during humification could be reflected in the geotechnical properties, specifically that mobilization of organic matter may result in a decrease in compressibility. Relationships between organic matter and sediment properties were also examined by Bennett et al. (1985) in two US continental margin areas. Although they found positive correlations between the plasticity-related properties and organic content in sediments from both the Mississippi Delta and the Central Atlantic margin, they concluded that the relationships were not strong enough to imply a significant interdependence, particularly at organic carbon levels below 5%. A series of experiments on sediment recovered from Chedabucto Bay, Nova Scotia, spiked with different amounts of organic matter, were conducted by Rashid and Brown (1975). Organic contents of 0%–4% were used. Again it was found that liquid and plastic limits increase with increasing organic content. Keller (1982) suggests that several index properties of marine sediments, including plasticity, correlate well with organic carbon at levels near 2%, but that strength characteristics may not be influenced until organic carbon contents reach 4%–5%. Booth and Dahl (1986) used a clayey silt from the Santa Barbara Basin (southern California) with varying amounts of organic matter to investigate the behavior of liquid and plastic limits, and grain-specific gravity. Liquid limit, plastic limit, and plasticity index all increased with increasing organic content over the range studied (0.57%–3.20% organic carbon) (Figures 6.52 and 6.53). Grain-specific gravity (i.e., specific gravity of solids, Gs) was shown to decrease (Figure 6.52). The above studies on the relationship between organic matter and sediments have been summarized by Booth and Dahl (1986) in Table 6.10. A review of this table shows that despite differences in materials and approaches, each of the investigations listed in Table 6.10 shows that both the liquid limit and the plastic limit increase with increasing organic content. In addition, it is also shown that the liquid limit increases at a slightly greater rate than the plastic limit. In addition, the relative rate of increase in the liquid and plastic limits is generally greater with an increase in organic content.

6.5.4 Compressibility/Compaction

The characteristics of organic soils in general are the following: (1) have low shear strength, (2) tend to be excessively compressible/ low compaction, and (3) tend to degrade with time. MacFarlane and Rutka (1962) evaluated the effect of muskeg on pavement structures. Figure 6.54 presented the K_o as a function of the over-consolidation ratio (OCR) for peat and clay. In Figure 6.55 they showed that the compression index as a function of natural void ratio plots as a linear relationship. This relationship can be described by the following equation.

$$C_c = 0.63(e_n - 1.0) \tag{6.25}$$

Franklin et al. (1973) studied the effect of organics content of soil compaction. The materials they studied were slightly organic soils, and mixtures of inorganic soil and

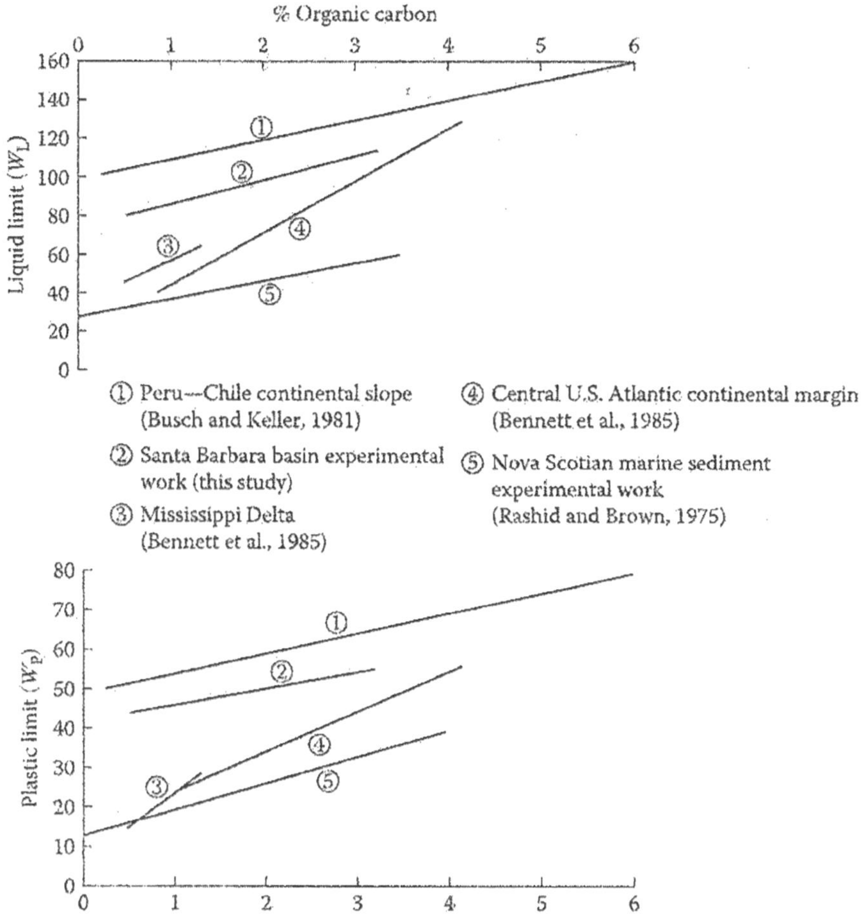

FIGURE 6.53 Relationship between organic matter and geotechnical properties in natural settings (After Booth and Dahl, 1986). Reprinted by permission of Taylor & Francis Group.

peat prepared to obtain various degrees of organic content. They found that the maximum dry unit weight decreases rapidly with organic contents of more than 8% to 10%. They also found that the optimum moisture content of the soil increases along with organic content.

Lancaster et al. (1996) have conducted studies on the effect of organics on soil compaction. They conducted their studies by mixing various percentages by mass of organic matter with a SP-SM soil (approximately 10% passing the No. 200 sieve and 6.8% organic content) and then running modified proctor compaction tests (ASTM D1557, Method A). Percent organic content was determined using a loss on ignition procedure. Results for a number of different organic materials (i.e., fibrous organic material: straw, shredded redwood bark, rice hulls; non-fibrous organic material: municipal sewage sludge) were used. A typical plot of dry unit weight versus water

TABLE 6.10
Qualitative Summary of Selected Studies on the Relationship between Organic Matter and Sediments or Soils

					Effect of Increase in Organic Matter		
Reference and Area	Texture	Mineralogy	Organic Matter and Type	Soil and Description	Liquid Limit (w_l)	Plastic Limit (w •.)	Plasticity Index (I~)
Odell et al. (1960)	Variable	Variable	Variable	Topsoils	Increase	Increase	Increase
Illinois							
Pusch (1973)	Variable	Variable	Variable	Sensitive clay soils	Increase	Increase	Increase
Sweden							
Rashib and Brown (1975)[a]	Constant	Constant	Organo-day complex	Marine	Increase	Increase	Increase
Chedabucto Bay, Nova Scotia							
Busch and Keller (1981)	Variable	Variable	Detrital, organo-clay	Fine-grained marine	Increase	Increase	Increase
Peru–Chile continental slope							
Bennetet al.	Variable	Variable	Unknown	Fine-grained marine	Increase	Increase	Increase
Mississippi Delta							
Central U.S. Atlantic continental margin	Variable	Variable	Unknown	Fine-grained marine	Increase	Increase	Increase
	Constant	Constant					
Santa Barbara Basin (off southern California)			Detrital, organo-day[b]	Fine-grained marine	Increase	Increase	Increase

Source: Booth and Dahl (1986). Reprinted with permission of Taylor & Francis.
[a] Experimental work quantity of organic matter altered.
[b] Assumed.

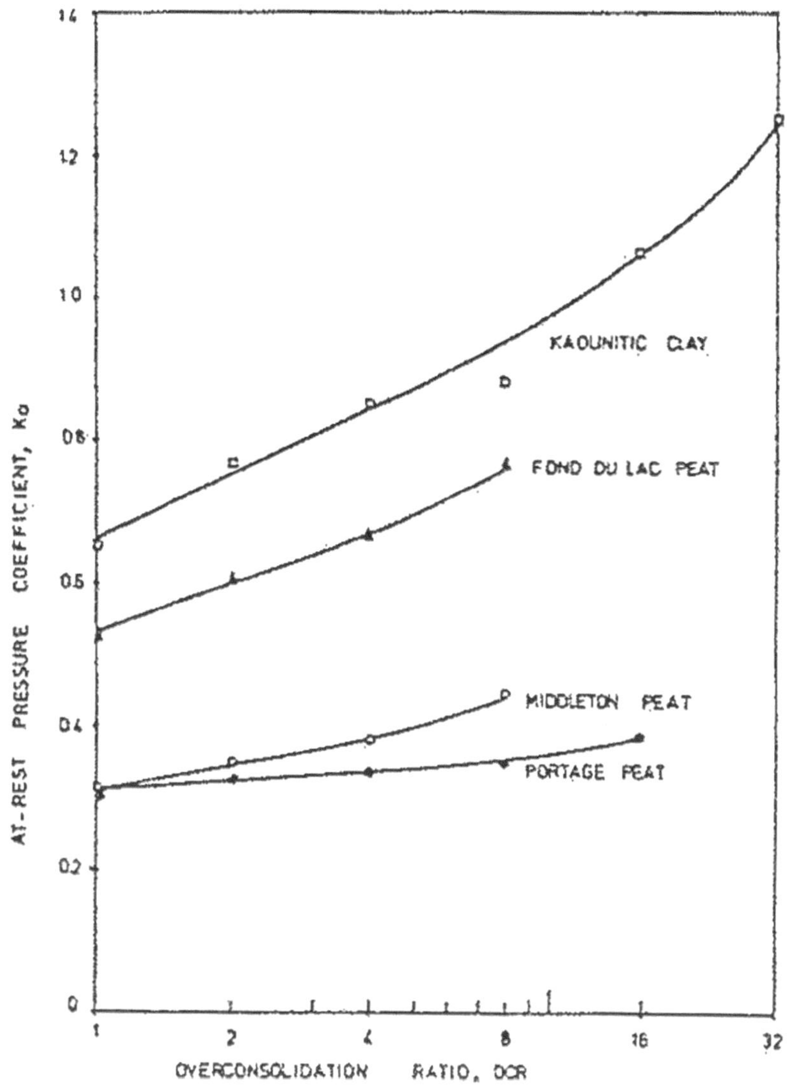

FIGURE 6.54 Earth pressure at rest as a function of OCR (Edil and Dhowian, 1981). Reprinted with permission of ASCE.

content shows that the maximum dry unit weight decreases while the optimum water content increases with increasing organic content. This occurs for both fibrous and non-fibrous material. Based on these limited results the actual relationship between the maximum dry unit weight and the true organic content for fibrous organic material is shown to be nonlinear in Figure 6.56. In contrast, the relationship between maximum dry unit weight and the true organic content for non-fibrous (sludge) organic material is linear.

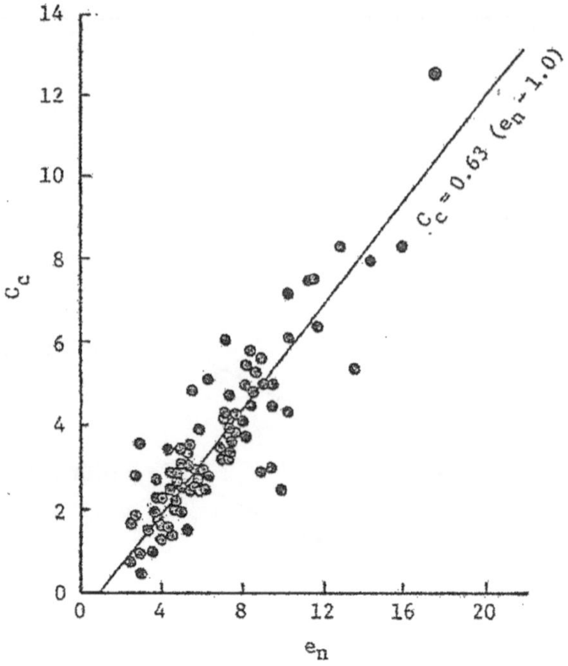

FIGURE 6.55 Consolidation index versus natural void ratio for muskeg material (Macfarlane and Rutka, 1962). Public domain.

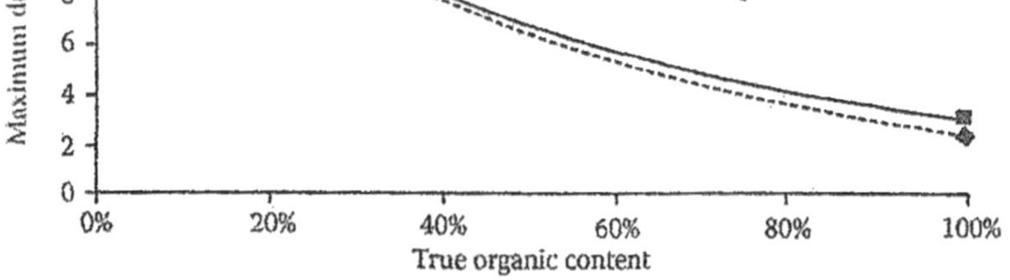

FIGURE 6.56 Comparison of maximum dry unit weight as a function of true organic content and material type (Lancaster et al., 1996). Reprinted with permission of the author.

The actual percent organic content of the soil as compacted incorporating both the initial and the added organic material can be calculated as follows:

$$\%ORG_s = \frac{(\%ORG_{sc})(\%ORG_{org}) + (100 - \%ORG_{sc})(\%ORG_{ss})}{100} \quad (6.26)$$

where

$\%ORG_{sc}$ is the percent of organic matter added
$\%ORG_{org}$ is the percent of the added organic soils that is actually organic
$\%ORG_{ss}$ is the percent organic matter of the base soil

The true organic content is then determined as presented in the following equation:

$$\%ORG_T = \frac{\%ORG_s}{\%ORG_s + 100} \quad (6.27)$$

A review of Figure 6.56 shows that for both fibrous and non-fibrous (i.e., sludge) organic material the maximum dry unit weight decreases with increasing organic content.

6.5.5 Permeability

Organic sediments have a high permeability when compared to other materials. The flow rate of water through the void space of a sediment is related to the fourth power of the pore radius (halving the pore radius drops the flow rate by a factor of 16). Therefore transmission of water is highly affected by compaction (i.e., decrease in void ratio).

6.5.6 Bioturbation

6.5.6.1 Introduction

Bioturbation is defined as a reworking of soils and sediments by plants and animals. These include burrowing ingestion and defecation of the sediment grains. Richards (1981) stated that benthic biologists believe that every grain of soil in the upper layers of the seabed has passed through the guts of some marine animals, which deposits a mucus membrane on particles.

A schematic diagram presenting a generalized burrow stratigraphy is shown in Figure 6.57.

Bioturbation over an area results in the mixing of sediments thus preventing the preservation of thin layers as shown in Figure 6.57. These layers are annual layers, thin turbidite layers or layers produced by contour currents alike. Modification of sediment structure by animal burrowing or surficial disturbance is recognized in core samples as distinct sections of infilled burrows or mottling. In addition, the effect of some microorganisms can be recognized by concentrations of micro-nodules or small black specks that occur commonly in non-laminated sediments. This effect occasionally is also found in the uppermost limits of graded silt-laminated zones. It is also common to find this effect occurring in brown, apparently oxidized clays.

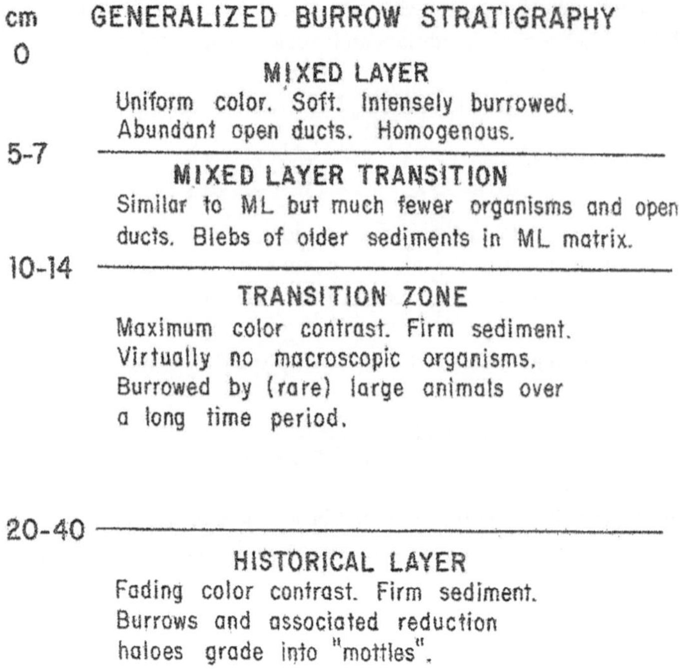

FIGURE 6.57 Generalized Burrow Stratigraphy (Berger, 1982).

When sampling in the NE Atlantic the presence of open burrows up to 2 m deep in some sediments has been observed. Two types of burrows were observed, one of up to 1.2 mm in diameter, the other about 5 mm in diameter. Similar types of burrows have been observed in both red clays from the Cape Basin and in the Pacific and in sediments from the NW African continental margin. How widespread such burrows are in the deep sea is, however, not known. Open burrows in deep sea sediments have a profound effects on the overall permeability and, consequently, on the possible flow rates through them in response to any excess pore pressures.

The categorization of burrowed media is based on geological records of trace fossils of biological activity (Gingras et al. 2015). This categorization depends on the distribution and characteristics of ichonofossils (i.e., size and diversity) of the assemblage. Ichologists have developed a bioturbation index (BI) in order to describe the degree to which sediments exhibit bioturbation (Taylor and Goldring, 1983).

This index classifies on a scale of 0–6 which is related to the abundance of fossil traces and their overlap, Table 6.11. The BI is related to the rate of sedimentation and duration of colonization.

- Highly and completely burrowed sediments.
  This is evidence of both a significant biomass and a slow sedimentation.
- Moderate to sparse bioturbation
  Evenly distributed trace fossils indicates a lower infauna biomass and higher sedimentation rate.

TABLE 6.11
Bioturbation Index

Bioturbation Index	Percent Bioturbated	Classification
0	0	No bioturbation
1	1–4	Sparse bioturbation, bedding distinct and few discrete traces or escape structures
2	5–30	Low bioturbation, bedding distinct, low trace density and escape structures often common
3	31–60	Moderate bioturbation, bedding boundaries sharp, traces discrete 3rd overlap rare
4	61–90	High bioturbation, bedding boundaries indistinct and high trace density with overlap common
5	91–99	Intense bioturbation, bedding completely disturbed (just visible), limited reworking and later burrows discrete
6	100	Complete bioturbation and sediment reworking because of repeated overprinting

Source: Taylor and Goldring (1983).

Highly to completely burrowed sediments are evidence of both a significant infaunal biomass and conditions of slow sediment accumulation. Moderate to sparse bioturbation, characterized by evenly distributed trace fossils indicates a lower infaunal biomass and higher sedimentation rate.

Sediments that appear to be either massive or homogeneous in texture can result from any of the following:

- Lack of sufficient grain size variation to define the sediment
- Sedimentation rate high enough that no grain-size segmentation occurs
- Mechanical mixing from soft sediment deformation during gravity pulls
- High degree of biogenic reworking.

The benthic invertebrates rework and modify sediments mechanically through activities such as burrowing, tube building, and deposit feeding. The results of these activities are evident in the variety of identifiable structures. These structures are both preserved in the sediment and at the current sediment water interface. The consolidation by fecal pelleting and casting, or tube construction and dispersive bioturbation are always common in sediments. The exception is where either sedimentation rates or physical erosion have obscured the structure or reworking.

The effects of bottom fauna on sediments at the head of Hudson Submarine Canyon has been documented using time-lapse photography.

Howard (1968) examined animals that are in the process of disturbing sediment by monitoring them using time lapse X-radiography. This was accomplished by using sediments contained in a narrow plexiglass aquarium connected to a continuously flowing seawater system. The burrowing animals to be studied are observed in the aquarium and the effect of their activities and the sediment is recorded by X-ray over a period of hours days and weeks

6.5.6.2 Shallow Water, Benthos and Sediments

In the shallow water of continental shelves and estuaries, the species of the benthic communities and their life histories are relatively well known. In the deep sea, or out past continental shelf depths, the Benthos and their habitats are not well documented. While survival of most shallow water living species depends on the conditions of the sediments, many forms are able to either markedly modify sediments, either by building protective habitation or through a variety a species – specific feeding activities intimately involving the sediments themselves. In shallow water many modifications known to be products of particular animals that are still in existence are easily recognized in the fossil record.

6.5.6.3 Bioturbated Deep Sea Sediment

A maximum burrowing depth of approximately 10–30 cm has been assumed based on studies of bioturbated layers of deep sea sediments. However, two cores obtained from the NE Atlantic within the Great Meteor East area, contained open burrows to a depth of 214 cm within a distal pelagic turbidite. Many of these burrows are less than 0.5 mm across and are only visible on fractured surfaces. Their effects on sediment mixing are probably minimal, but they provide a network of open vertical channels that can increase the permeability. The effect of the burrows changes the calculated permeability from the equivalent of a clay to that of a coarse sand. The delicate structure of these burrows may cause their closure during handling. If this happens it would require the effect of pore water fluxes to be studied in situ.

It has been shown that, for a mass of sediment containing smooth open parallel tubes (assuming Darcian and Poiseuille flow in the sediment and tubes respectively), the absolute permeability in the direction of the tubes, K (cm^2), is given by the following (Schultheiss and Thomson, 1984):

$$K = K_s\left(1 - nr^2\right) + nr^4/8 \quad (6.28)$$

Where

K_s is the absolute permeability of the intact sediment (cm^2)
n is the number of tubes per square centimeter
r is the radius of the tubes (cm)

The influence of open burrows on the permeability of deep sea clays using the above equation for a range of values of r and n is illustrated in Figure 6.58. K_s becomes negligible in the above equation when the values fall below 10^{-10} cm^2. Most deep sea sediments have permeabilities that range between 10^{-10} and 10^{-12}cm^2 so the increase by two orders of magnitude produced by the open burrows (where $n = 0.5$ and $r = 0.2$ mm) becomes significant. In red clays, with typical permeabilities of 10^{-12}cm^2, the effect would be to increase the permeability by four orders of magnitude.

Laboratory measurements on cores give K_s values around 10^{-11}cm^2 for the interval below the open burrow. In the burrowed interval, however, values of 10^{-10}cm^2 were recorded. The reason for values much lower than those predicted in Figure 6.59 may be two-fold. One, permeability measurements at low effective stresses are often inaccurate due to sample sealing problems. Two, the delicate burrows may close, at least partially, during sampling (Schultheiss and Thomson, 1984).

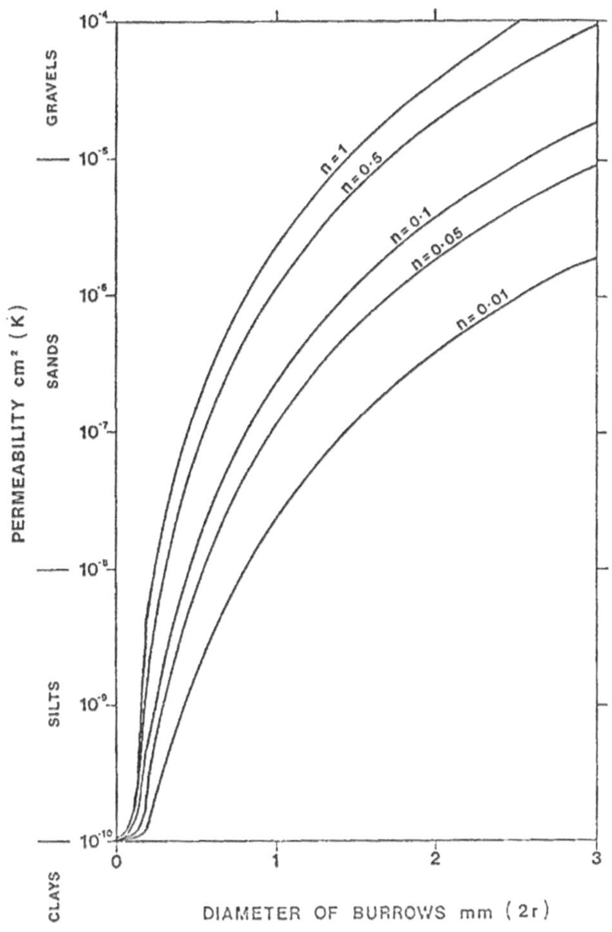

FIGURE 6.58 Permeability as a function of diameter of burrows and number (Schultheiss and Gunn, 1985).

Highly to completely burrowed sediments are evidence of both a significant infaunal biomass and conditions of slow sediment accumulation. Moderate to sparse bioturbation, characterized by evenly distributed trace fossils indicates a lower infaunal biomass and higher sedimentation rate.

REFERENCES

Angemeer, J. and MCNeilian, T.W. (1982). "Subsurface variability—The key to investigation of Coral Atoll," in K. Demars and R.C. Chaney (Eds.), *Geotechnical Properties, Behavior, and Performance of Calcareous Soils*, STP 777, American Society for Testing and Materials, Philadelphia, PA, pp. 36–53.

Arman, A. (1970). "Engineering classification of organic soils," Highway Research Record No. 310, pp. 75–80.

ASTM. (2014). *D653-Standard Terminology Related to Soil, Rock, and Contained Fluids*. American Society for Testing Materials, Philadelphia, PA, 46pp.

Attwooll, W.J., Fujioka, M.R., and Kittridge, J.C. (1985). "Laboratory and in-situ strength testing of marine soils in Iran and Indonesia," in R.C. Chaney and K.R. Demars (Eds.), *Strength Testing of Marine Sediments: Laboratory and In-Situ Measurements*, ASTM STP 883, American Society for Testing and Materials, Philadelphia, PA, pp. 440–453.

Bennett, R.H., Lehman, L., Hulbert, M.H., Harvey, G.R., Bush, S.A., Forde, E.B., Crews, P., and Sawyer, W.B. (1985). "Interrelationships of organic carbon and submarine sediment geotechnical properties," *Marine Geotechnology*, 6(1): 61–98.

Berger, W.H. (1982). "The benthic interface of deep sea carbonates: A three-tiered sequence controlled by depth of deposition," in K.A. Fanning and F.T. Manheim (Eds.), *Dynamic Environment of the Ocean Floor*, Lexington Books, Lexington, MA, pp. 95–116.

Beringen, E.L., Kolk, H.J., and Windle, D. (1982). "Cone penetration and laboratory testing in marine calcareous sediments," in K.R. Demars and R.C. Chaney (Eds.), *Geotechnical Properties, Behavior and Performance of Calcareous Soils*, ASTM STP 777, American Society for Testing and Materials, Philadelphia, PA, pp. 179–209.

Bjerrum, L. (1967). "Engineering geology of Norwegian normally-consolidated marine clays as related to settlement of buildings," *Geotechnique*, 17(2): 81–118.

Booth, J.S. and Dahl, A.G. (1986). "A note on the relationships between organic matter and some geotechnical properties of a marine sediment," *Marine Geotechnology*, 6(3): 281–287.

Bryant, W.R., Deflanche, A.P., and Trabant, P.H. (1974). "Consolidation of marine clays and carbonates," in A.L. Inderbitzen (Ed.), *Deep-Sea Sediments: Physical and Mechanical Properties*, Plenum Press, New York, pp. 209–244.

Bryant, W.R., Hottman, W., and Trabant, P. (1975). "Permeability of unconsolidated and consolidated marine sediments, gulf of Mexico," *Marine Geotechnology*, 1(1): 1–14.

Busch, W.H., and Keller, G.H. (1981). "The physical properties of Peru–Chile continental margin sediment – The influence of coastal upwelling on sediment properties," *Journal of Sedimentary Petrology*, 51: 705–719.

Casagrande, A. (1932). "The structure of clay and its importance in foundation engineering," *Boston Society of Civil Engineers*, 19(4): 168.

Chan, H.E., and Kenney, T.C. (1973). "Laboratory investigation of the permeability ratio of New Liskeard vared soil," *Canadian Geotechnical Journal*, 10: 453–472.

Chaney, R.C., and Almagor, G. (2016). *Seafloor Processes and Geotechnology*, CRC Press, Taylor & Francis Group, Boca Raton, FL, 558pp.

Chaney, R., and Fang, H.Y. (1986) "Static and dynamic properties of marine sediments: A state of the art," in R.C. Chaney and H.Y. Fang (Eds.), *Marine Ceotechnogy and Nearshore/Offshore Structures*, STP 923, ASTM Press, Philadelphia, PA, pp. 74–11.

Chassefiere, B., and Monaco, A. (1983). "On the use of Atterberg limits on marine soils," *Marine Geotechnology*, 5(2): 153–179.

Cox, J.B. (1970). "Shear strength characteristics of the recent marine clays in south east Asia," *Journal of the Southeast Asian Society of Soil Engineering*, 1: 1–28.

Clukey, E.C., and Silva, A.J. (1982). "Permeability of deep-sea clays: Northwestern Atlantic," *Marine Geotechnology*, 5(1): 1–26.

Darcy, A. (1856). *Les Fontaines Publiques de la Ville de Dijon*, Dalmont, Paris, France.

Datta, M., Gulhati, S.K., and Rao, G.V. (1979). "Crushing of calcareous sands during shear," in *Proceedings, Offshore Technology Conference*, vol. 3, Houston, pp. 1459–1467.

Davie, J.R., Fenske, C.W., and Serocki, S.T. (1978). "Geotechnical properties of deep continental margin soils," *Marine Geotechnology*, 3(1): 85–116.

Demars, K.R. (1982). "Uinque engineering properties and compression behavior of deep-sea calcareous sediments," in K.R. Demars and R.C. Chaney (Eds.), *Geotechnical*

Properties: Behavior and Performance of Calcareous Soils, ASTM 777, American Society for Testing and Materials, Philadelphia, PA, pp. 97–112.

Demars, K.R., and Anderson, D.G. (1971). "Environmental factors affecting the emplacement of seafloor installations," Technical Report TR761, Naval Civil Engineering Laboratory, Port Hueneme, CA, March, 114pp.

Demars, K.R., Nacci, V.A., Kelly, W.E., and Wang, M.C. (1976, January 1). "Carbonate content: An index property for ocean sediments," *Offshore Technology Conference*, doi:10.4043/2627-MS

Edil, T.B., and Dhowian, A.W. (1981). "At-rest lateral pressure of peat soils," *Journal of the Geotechnical Engineering Division, Proceedings ASCE*, 107(GT2): 201–216.

Farrar, D.M., and Coleman, J.D. (1967). "The correlation of surface area with other properties of nineteen British clay soils," *Journal of Soil Science*, 18(1): 118–124.

Franklin, A.F., Orozco, L.F., and Semrau, R. (1973). "Compaction of slightly organic soils," *J Soil Mech Div-Asce*, 99: 541–557.

Geotechnical Consortium. (1985). Geotechnical properties of northwest pacific pelagic clays: Deep seadrilling project leg 86, Hole 576A, Part VI Physical Properties and Geotermal Studies, Vol. LXXXVI, NSF: 723–758.

Gibson, R.E. (1958). "The progress of consolidation in a clay layer increasing in thickness with time," *Geotechnique*, 8: 171–182.

Gingras, M.K., Pemberton, S.G., and Smith, M. (2015). "Bioturbation: Reworking sediments for better or worse," *Oilfield Review*, 26(4): 46–50.

Gray, D.H., and Mitchell, J.K. (1967). "Fundamental aspects of electric osmosis in soils," *Journal of the Soil Mechanics and Foundations Division*, 93(6): 209–236.

Gross, M.G. (1967). *Oceanography: A View of the Earth*, United States Department of Energy, N.P., Web.

Gupta, K.P., and Swartzendruber, D. (1962). *Flow Associated Reduction in the Hydraulic Conductivity of Quartz Sand. Soil Science Society of America Proceedings. Soil Science of America*, Madison, WI, pp. 6–10.

Herrmann, H.G., Raecke, D.A., and Albertson, N.D. (1972). "Selection of practical seafloor foundation systems," Technical Report TR761, Naval Civil Engineering Laboratory, Port Hueneme, March, 114pp.

Howard, J.D. (1968). "X-ray radiography for examination of burrowing in sediments by marine invertebrate organisms," *Sedimentology*, 11: 249–258.

IOS (Institute of Oceanographic Sciences). (1978). "Oceanography related to waste disposal," Report No. 77, Wormley, Godalming, Surrey.

Keller, G.H. (1982). "Organic matter and the geotechnical properties of submarine sediments," *Geo-Marine Letters*, 2: 191–196.

Hamilton, E.L. (1976). "Variation of density and porosity with depth in deep-sea sediments," *Journal of Sedimentary Petrology*, 46, 280–300.

Holtz, R.D., and Kovacs, W.D. (1981). *An Introduction to Geotechnical Engineering*, Prentice Hall, Englewood Cliffs, NJ.

Horn, D.R., Delaach, M.N., and Horn, B.M. (1976). "Physical properties of sedimentary provinces, North Pacific and North American Ocean," in A.L. Inderbitzen (Ed.), *Deep-Sea Sediments*, Plenum Press, New York, pp. 417–441.

Hunt, J.M. (1979). *Petroleum Geochemistry and Geology*, W.H. Freeman, San Francisco, CA.

Lancaster, J.K., Waco, R., Towle, J., and Chaney, R.C. (1996). "The effect of organic content on soil compaction," in *Proceedings 3rd Internatinal Symposium on Environmental Geotechnology*, vol. 1, San Diego, CA, pp. 152–161.

Lee, H. J. (1982). "Bulk density and shear strength of several deep-sea calcareous sediments," in: K.R. Demars and R.C. Chaney (Eds.), *Geotechnical Properties: Behavior and Performance of Calcareous Soils*, ASTM 777, American Society for Testing and Materials, Philadelphia, PA, pp. 36–51.

Lupini, J.F., Skinner, A.E., and Vaughan, P.R. (1981). "The drained residual strength of cohesive soil," *Geotechnique*, 31(2): 181–213.
MacFarlane, I.C., and Rutka, A. (1962). "An evaluation of pavement performance over muskeg in Northern Ontario," *Highway Research Board Bulletin*, 316, 32–43.
Mesri, G. (1973). "Coefficient of secondary compression," *Journal of the Soil Mechanics and Foundations Division, ASCE*, 99(1): 123–137.
Mesri, G. and Godlewski, P.M. (1977). "Time and stress-compressibility interrelationship," *Journal of the Geotechnical Engineering Division*, 103(5): 417–430.
Michaels, A.S., and Lin, C.S. (1954). "The permeability of kaolinite," *Industrial and Engineering Chemistry*, 46(6): 1239–1246.
Miller, R.H., Overman, A.R., and Peverly, J.H. (1969). "The absence of threshold gradients in clay-water systems," *Soil Science Society of America Proceedings*, 33(2): 183–187.
Mitchell, J.K. (1976). *Fundamentals of Soil Behavior*, Wiley, New York.
Mitchell, J.K., and Younger, J.S. (1967). Abnormalities in hydraulic flow through fine-grained soils. *ASTM Special Testing Publication*, 417: 106–141.
Mitchell, J.K. (1993). *Fundamentals of Soil Behavior,* 2nd edition. Wiley, New York.
Moum, J., and Rosenquvist, I.T. (1961). "The mechanical properties of montmorillonitic and Illitic clays related to the electrolytes of the pore water," *Proceedings of the 5th International Conference on Soil Mechanics and Foundation Engineering*, 1: 263–267.
Mullins C.E., MacLeod D.A., Northcote K.H., Tisdall J.M., and Young I.M. (1990). "Hardsetting soils: Behavior, occurrence, and management," in R. Lal, and B.A. Stewart (Eds.) *Advances in Soil Science*, vol. 11. Springer, New York, NY.
Nacci, V.A., Kelly, W. E., Wang, M.C., and Demars, K.R. (1974). "Strength and stress-strain characteristics of cemented deep-sea sediments," in A.I. Inderbitzen (Ed.), *Deeo-Sea Sediments: Physical and Mechanical Properties*, Plenum Press, New York.
Nakase, A., Kamei, T., and Kusakabe, O. (1988). "Constitutive parameters estimated by plasticity index," *Journal of Geotechnical Engineering, ASCE*, 114 (GT4): 844–858.
Nauroy, J.-F., and Tirant, P. (1982). "Behaviour of marine carbonate sediments," *Research Gate*, 37: 149–155.
Noorany, I. (1983). *Classification of Marine Sediments*, Soil Mechanics Series, San Diego State University, San Diego, CA, 27pp.
Odell, K.T., Thornburn, T.H., and McKenzie, L. (1960). "Relationships of Atterberg limit to some other properties of Illinois soils.," *Proceedings of the Soil Science Society of America*, 24(5): 297–300.
Ohira, Y. (1977). *Engineering Problems of Organic Soils in Japan*, Japanese Society of SMFE, Tokyo, pp. 19-34.
Oikawa, H., and Igarashi, M. (1997). "A method for predicting e-log p curve and log cv – log p curve of a peat from its natural water content," in B.K. Huat and H.M. Bahia (Eds.), *Proceedings of the Recent Advances in Soft Soil Engineering*, Kuching, Sarawak, Malaysia.
Olsen, H.W. (1965). "Deviations from Darcy's Law in saturated clays," *Soil Science Society of America, Proceedings*, 29: 135–140.
Olsen, H.W. (1969). "Simultaneous fluxes of liquid and charge in saturated kaolinite," *Soil Science Society of America, Proceedings*, 33(3): 338–344.
Poulos, H.G. (1984). "Cyclic degradation of pile performance in calcareous soils," *Analysis and Design of Pile Foundations, ASCE*, 99–118.
Poulos, H.G. (1988). *Marine Geotechnics*, Chap. 3, Unwin Hyman, London.
Poulos, H.G., Uesugi, M., and Young, G.S. (1982). "Strength and deformation properties of Bass Strait carbonate sands," *Geotechnical Engineering, Southeast Asian Geotechnical Society*, 11: 189–211.

Pusch, R. (1973). "Influence of organic matter on the geotechnical properties of clays," National Swedish Institute for Building Research, Summaries, D6, Chalmers Institute of Technology, Department of Geotechnics, Gothenburg, Sweden.
Rashid, M.A., and Brown, J.D. (1975). "Influence of marine organic compounds on the engineering properties of a remolded sediment," *Engineering Geology*, 9: 141–154.
Raymond, G.P., and Wahls, H.E. (1976). *Estimating One Dimensional Consolidation, including Secondary Compression of Clay Loaded from Overconsolidated to Normally Consolidated State.* Special Report 163. TRB, Washington, DC, pp. 17–23.
Richards, A.F. (1981). Personal communication.
Schultheiss, P.J. and Thomson, J. (1984). "Disposal in sea-bed geological formations—Properties of ocean sediments in relation to the disposal of radioactive waste," Commission of the European Communities, EUR 8962.
Schultheiss, P.J., and Gunn, D.E. (1985). "The permeability and consolidation of deep-sea sediments," Report No. 201, Institute of Oceanographic Sciences, Wormley, UK, 94pp.
Seed, H.B., Woodward, R.J., and Lundgren, R. (1964). Clay mineralogical aspects of the Atterberg Limits. *Journal of Soil Mechanics and Foundation Division*, 10(4): 107–131.
Shepard, L.E. (1986). *Geotechnical Property Characteristics of Nares Abyssal Plain Sediment: A Summary Report*, Geological studies of the Southern Nares Abyssal Plain, Western North Atlantic, Rijks Geologische Dienst, Internal Report.
Silva, A.J., Hetherman, J.R., and Calnan, D.I. (1981). "Low-gradient permeability testing of fine-grained marine sediments," in T.F. Zimmie and C.O. Riggs (Eds.), *Permeability and Groundwater Contaminant Transport*, ASTM STP 746, American Society for Testing and Materials, Philadelphia, PA, pp. 121–136.
Skempton, A.W. (1957). "Discussion: Further data on the c/p ratio innormally consolidated clays," *Proceedings of the Institution of Civil Engineers*, 7: 305–307.
Soderblom, R. (1969). "Salt in Swedish clays and its importance for quick clay formation," *Swedish Geotechnical Proceedings*, 22: 63.
Sverdrup, H.U., Johnson, M.W., and Fleming, R.H. (1942). *The Oceans: Their Physics, Chemistry, and General Biology*, Prentice-Hall, Englewood Cliffs, NJ, 1087pp.
Taylor, A.M., and Goldring, R. (1983). "Description and analysis of bioturbation and ichnofabric," *Journal of the Geological Society*, 150(January): 141–148.
Valent, P.J., Altschaeffl, A.G., and Lee, H.J. (1982). "Geotechnical properties of two calcareous oozes, geotechnical properties, behavior, and performance of calcareous soils," in K.R. Demars and R.C. Chaney (Eds.), American Society for Testing and Materials, Philadelphia, PA, pp. 79–95.
Yin, J.H., H.J. Liao, Zhon, C., and Cheng, C.M. (2003). "Estimation of marine soil parameters for preliminary analysis of geotechnical structures in the Taiwan Strait Connection Project," *Journal of Marine Georesources and Geotechnology*, 21(3–4): 167–182.
White, H.E., and Walton, S.F. (1937). "Particle packing and particle shape," *Journal of the American Ceramic Society*, 20(1): 155–166.

7 Behavior of Marine Sediments under Sustained, Dynamic and Cyclic Loading

7.1 INTRODUCTION

The behavior of sediments under load is dependent upon a number of factors. These factors include type of loading (i.e., long-term sustained load application, dynamic loading, cyclic loading), type of pore pressure response (drained or undrained) and the permeability, presence of cohesion or cementation between particles.

7.2 SUSTAINED LOADING

Creep is long-term continuing deformation due to sustained deviatoric stress

$$\sigma_d = \sigma_1 - \sigma_3 \tag{7.1}$$

σ conditions that occurs as a function of time after dissipation of consolidation excess pore pressures. Thus, creep behavior of sediments is a function of the type of sediment, its physical properties, stress–strain history, and time. Mitchell (1976) distinguished between creep and secondary compression by noting that the former refers to "time-dependent shear strains that develop at a rate controlled by the viscous resistance of the soil structure." Secondary compression in contrast involves the "specific case of volumetric strain that follows primary consolidation and is controlled in rate by the viscous resistance of the soil skeleton as opposed to hydrodynamic." By contrast, the stress relaxation is the long term decrease in strength that occurs as a function of time under a sustained strain level. In this section, only creep behavior will be discussed. The time-dependent behavior of both clay and sand, and the methods to characterize their response to load have been presented by a number of authors. A few of these authors are as follows: Augustesen et al. (2004), Liingaard et al. (2004), Vyalov (1986), and Lade et al. (2009). The primary point that these authors emphasize is that clay and sand respond differently with respect to time. A schematic illustrating this difference is presented in Figure 7.1 (Lade et al., 2009). A review of Figure 7.1a shows that strain rate has an important influence.

If the strain rate is changed during loading, then permanent effects in the resulting overall strain occur (Figure 7.1b). By contrast, different strain rates produce similar stress–strain curves for sand (Figure 7.1c). This behavior of sand has been reported

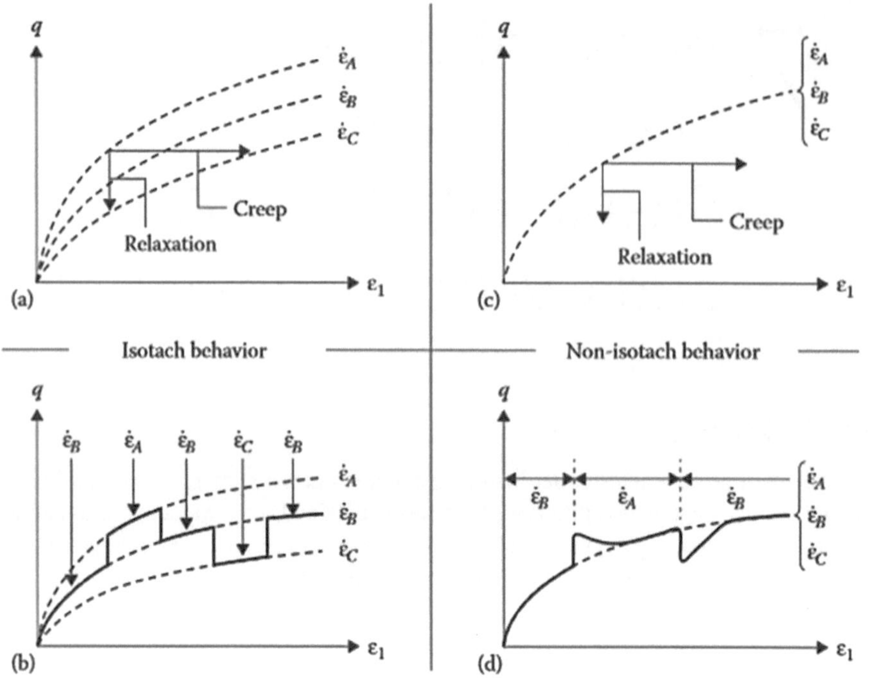

FIGURE 7.1 Isotach behavior observed in clay for (a) creep and relaxation and (b) stepwise change in strain rate. Nonisotach behavior observed in sand for (c) creep and relaxation and (d) stepwise change in strain rate (From Lade et al., 2009). Reprinted with permission of ASCE.

by a number of investigators: Lade et al. (2009), Matsushita et al. (1999), Tatsuoka et al. (2000, 2002, 2006), Di Benedetto et al. (2002), Kuwano and Jardine (2002), AnhDan et al. (2006), and Kiyota and Tatsuoka (2006), to name a few. If the strain rate is changed during loading a sand material, only temporary changes occur (Figure 7.1d). In the following, discussion will be limited to silts and clays. To visualize the effect of time on the behavior of sediments, uniaxial compression creep tests can be run on cylindrical specimens, subjected to a constant uniaxial stress. When a step load is applied on the specimen, it responds by an instantaneous deformation followed by creep. A typical creep curve (strain [ε] vs. time) resulting from such a step loading is shown in Figure 7.2a. This curve consists of three stages: stage I is the primary stage, followed by the apparent secondary stage II and finally by the tertiary stage III. The related curve of strain rate ($\dot{\varepsilon}_{\min}$) versus time (t) curve is presented in Figure 7.2b. The creep curve is reflected in the creep rate ($\dot{\varepsilon}$). Initially, the creep rate ($\dot{\varepsilon}$) decreases (i.e., primary creep, primary stage I); then, it is essentially constant (i.e., secondary or steady-state creep, apparent secondary stage II). This corresponds to a minimum strain rate (εmin•). An increasing strain (tertiary creep, tertiary stage III) follows, which eventually leads to ultimate failure of the specimen. The steady-state portion of the creep curve is usually reduced to an inflection point (m). At low stress levels (SLs), primary creep appears to dominate. Figure 7.2c presents a

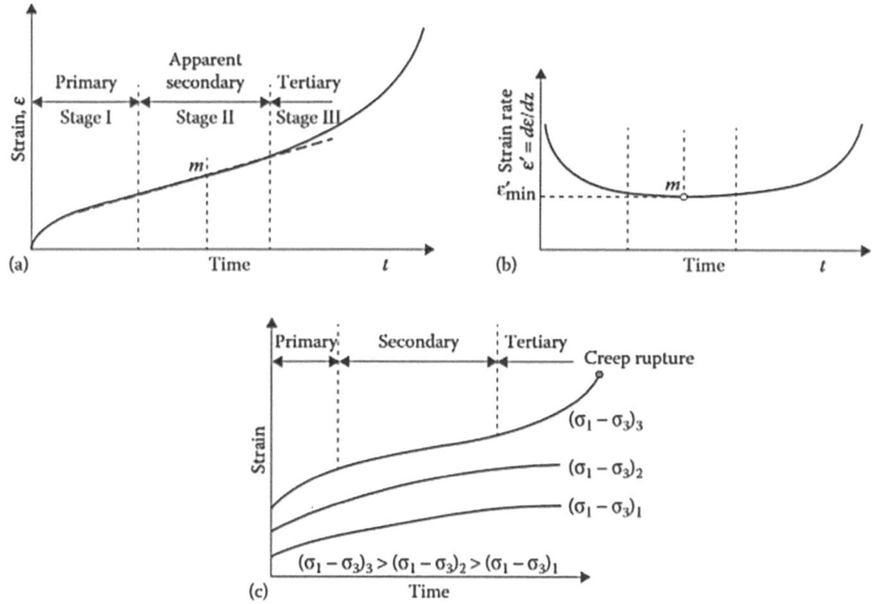

FIGURE 7.2 Strain and strain rate versus time (Chaney and Almagor, 2016). Reprinted with permission of Taylor & Francis.

schematic of the effect of an increasing level of deviatoric stress ($\sigma_d = \sigma_1 - \sigma_3$) on the level of strain as a function of time. As the applied deviatoric stress ($\sigma_d = \sigma_1 - \sigma_3$) is increased, the ultimate deviatoric stress strain as a function of time increases until creep rupture occurs. The proportion of the total curve each stage represents is dependent on a number of factors. These factors are the material type, the SL used during the test, and to a lesser degree the specimen shape and test conditions. The relationship between steady-state creep rate (SSCR) and a number of factors is shown in Table 7.1. A review of this table shows that as plasticity index, % creep, and activity increases, the SSCR increases. By contrast, as the time increases for a drained test, the SSCR decreases. For an undrained test, the SSCR decreases until critical then increases. If the ratio of deviator stress/initial strength increases, then the SSCR increases. A plot of the logarithm of strain rate decreases as a function of the logarithm of time and is linear. The same behavior has been observed for undisturbed and remolded wet or dry clay, NC and OC clay, and sand (Singh and Mitchell, 1968). In general, different soil types exhibit varying amounts of time-dependent deformations and stress variations with time. These variations are also exhibited by their secondary compression and creep characteristics. The amount of deformation that might be expected from a given load is of great practical importance. To estimate the amount of deformation that can potentially occur, short-term laboratory tests are typically run to determine a soil's creep properties. The significance of using results from short-term tests to extrapolate to actual project design lifetimes is unknown. A number of mathematical expressions to describe the time deformation behavior have been developed by various authors to model the laboratory behavior. Singh (1966) has

TABLE 7.1
Factors Affecting the Steady-State Creep

Factors Increasing	SSCR[c]
PI	Increases
% Creep	Increases
Activity	Increases
Time[a]	Decreases
Time[b]	Decreases until critical, then increases
Deviator stress/Initial strength	Increases

Source: Chaney and Almagor (2016). Reprinted with permission of Taylor & Francis.
[a] Drained
[b] Undrained
[c] SSCR, steady-state-creep-rate

broken these various relationships into three categories: the viscoelastic (i.e., rheological) model approach, the rate process approach, and the empirical approach. In the following, each of these approaches will be discussed.

7.2.1 Viscoelastic Model Approach

Viscoelasticity is the property of materials that exhibit both viscous and elastic characteristics when undergoing deformation. To model this material behavior, viscoelasticity utilizes spring constants (E_1 and E_2), dashpots (viscosity, v), and St. Venant sliders (a slider to account for non-recoverable deformation) elements. The properties of these elements may be selected to cover a wide range of elastic and time-dependent viscous behavior. Viscoelastic models can be divided into both the number of elements employed and whether the elements are in a series or parallel arrangements. These elements may be linear or nonlinear and are combined as necessary for the model to describe the behavior of the sediment under study. These models describe short-term behavior reasonably well, but tend to not yield reliable predictions of deformation for extended time periods. A number of viscoelastic (i.e., rheological) models have been proposed to model steady-state creep in soils. A selection of four of these models is presented in Figure 7.3. A model incorporating spring constants, E_1 and E_2; a slider element of resistance to; and a dash-pot with viscosity, v, was proposed by Murayama and Shibata (1964), which is shown in Figure 7.3a. The time-dependent deformation is controlled by the slider element τ_0. Deformation will only occur for applied stresses in excess of τ_0. Schiffman (1959) proposed a model to predict steady-state creep that incorporates a Maxwell element in series with a Kelvin element (Figure 7.3b).Tests on many soils have failed to show the evidence of steady-state creep conditions even though it is widely used in engineering approximations. Christensen and Wu (1964) presented a three-element model that requires the soil to reach a maximum ultimate strain under a given load (Figure 7.3c). This model incorporates a dashpot that is controlled by a hyperbolic law. It is difficult to verify experimentally because strain rates approaching zero are difficult and depend largely on the sensitivity of the detection devices. A three-element rheological model

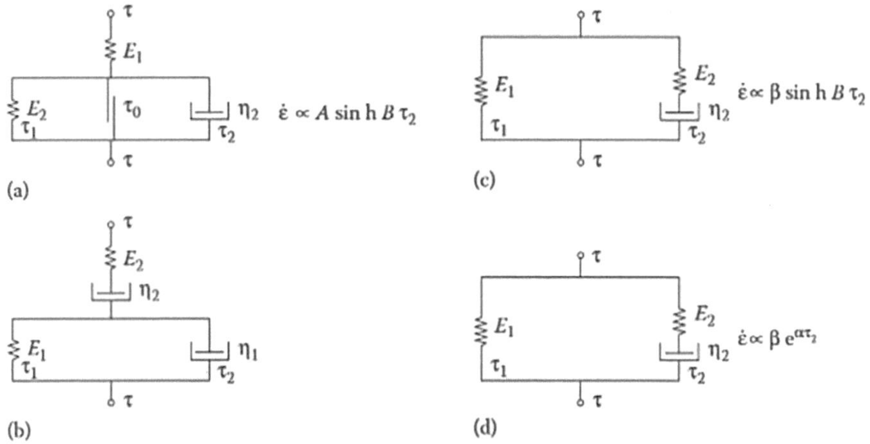

FIGURE 7.3 Viscoelastic models to simulate soil creep. Equations to the right of model govern dashpot behavior. (a) Murayama and Shibata (1964); (b) Schiffman (1959); (c) Christensen and Wu (1964); (d) Singh (1966) (From Hirst, 1968a).

was proposed by Singh (1966) (Figure 7.3d). This model is similar to Christensen and Wu's, with the exception that the dashpot obeys an exponential rather than a hyperbolic law. This short summary has shown that viscoelastic models are capable, for engineering purposes, of describing the time–deformation relationship of soils over a given time interval and a range of applied stresses. In general, the method does not offer a correct mathematical description of soil behavior over all times and all stresses.

7.2.2 Rate Process Approach

Time-dependent behavior of soil occurs during deformation and shear failure can be described as a rate process. The theory of absolute reaction rates was developed by Glasstone, Laidler, and Eyring (1941). Application of this theory to the problem of soil creep of soils has been studied by a number of authors: Murayama and Shibata (1964); Christensen and Wu (1964); Mitchell (1964); Mitchell, Campanella, and Singh (1968); Feda (1989); and Kuhn and Mitchell (1992). The rate process theory is based on assuming that particles participating in a time-dependent deformation process are restricted from movement relative to each other. The restraint is usually thought of as an amount of potential energy (i.e., differential height, cohesion, friction, etc.) that has to be expended before displacement can occur. Therefore, for an element in equilibrium that is located at point A to displace to another point B that requires energy/work (i.e., activation energy), DF needs to be expended. To continue this movement requires the next barrier (i.e., the amount of potential energy that has to be expended) to be surmounted. The amount of energy DF required could involve one of three separate cases: (1) DF remains the same, (2) DF increases, or (3) DF decreases. In soils, typically DF is assumed to remain constant.

The following equation to describe the strain rate (ε') of soils at any time has been presented by Hirst (1968b)

$$\varepsilon' = 2X\frac{\kappa T}{h}\exp\left(\frac{-\Delta F}{RT}\right)\sinh\left(\frac{f\lambda}{2\kappa T}\right) \tag{7.2}$$

where

ε' is the strain rate at any time t
X is the a constant depending upon the soil structure at time t
k is Boltzmann's constant
h is Planck's constant
R is the gas constant
T is the absolute temperature
f is the average shear force acting across each interparticle bond
λ is the an average distance between successive equilibrium positions in the interparticle contact zone
DF is the free energy of activation

Mitchell (1993) indicates that there is no rigorous proof of the statistical mechanics formulation of the rate process theory, but it does tend to describe the behavior of many real systems. The method, although interesting, is not normally used in practice.

7.2.3 Empirical Approach

An empirical method is the third approach used to describe time-dependent deformations in soils. This approach involves modeling the behavior of the material under controlled conditions. Attempts are then made to develop mathematical relationships between the system parameters to obtain equations that both describe and predict the material behavior.

The variation of axial strain rate of creep as a function of time is presented in Figure 7.4. A review shows that for SL less than or equal to 0.83, the relationship of log $\dot{\varepsilon}_a$ versus log t is linear. The SL is defined as the ratio of the imposed deviator stress and the deviator stress at failure in the conventional triaxial test. A number of authors (Singh and Mitchell, 1968; Vaid and Campanella, 1977; Tavenas et al., 1978; Zhu et al., 1999) have shown that for low SL, the lines in Figure 7.4 are nearly parallel for creep tests. This indicates that the slope of the lines (m) is constant. A close review of this figure indicates that the individual lines intersect at approximately 10 min but have different slopes. A review of this figure shows that as the SL becomes smaller, the larger the value of m. A schematic plot of log $\dot{\varepsilon}_a$ versus the principal stress difference (D) as a function of time is presented in Figure 7.5.

A review of Figure 7.5 shows that at a time of $t = 1$ min after the start of creep, the slope of the linear part of the curve can be projected back to the ordinate axis, which gives a value for a parameter A. The strain rate based on the curves in Figure 7.5 has been presented by Singh and Mitchell (1968) in the following equation:

$$\dot{\epsilon} = Ae^{\alpha q}\left(\frac{t_1}{t}\right)^m \tag{7.3}$$

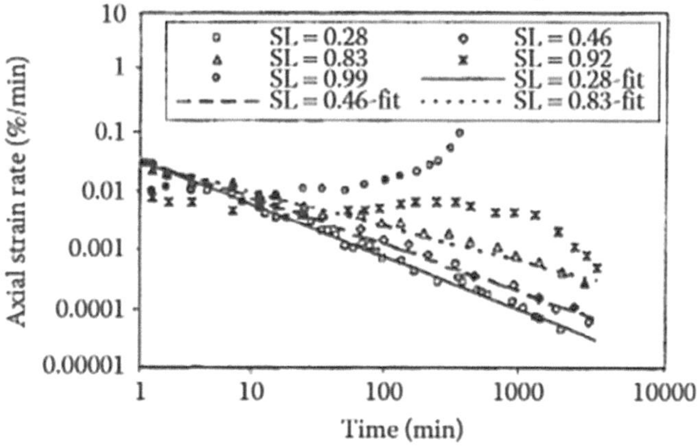

FIGURE 7.4 Log axial strain rate versus log time (From Zhu et al., 1999.) Reprinted with permission of ASTM.

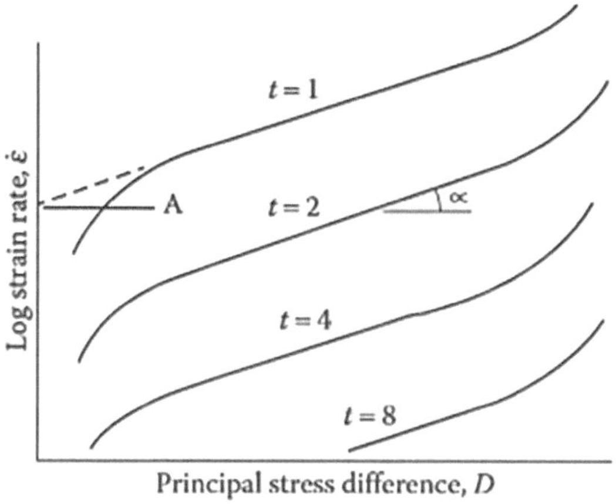

FIGURE 7.5 Log strain rate versus principal stress difference.

where

A, a, and m are material parameters
q is the deviator stress
t is the unit time (i.e., 1 min)

Parameters a and m have been shown to be functions of both the current stress history. This is based on creep behavior from multistage testing (Zhu et al., 1999). Parameter A was shown to be a function of clay type (Hirst, 1968b).

Integrating Equation 7.3 gives an equation for strain (ϵ). This equation predicts uniaxial creep behavior for both drained and undrained conditions. A minimum of two creep tests is required to determine the values of the parameters *A*, *m*, and *a*.

$$\epsilon = \frac{At_1}{(1-m)} e^{\alpha D} \left(\frac{t_1}{t} \right)^{(1-m)} \tag{7.4}$$

7.3 DYNAMIC AND CYCLIC LOADING

The primary difficulty with wave or seismic loading on seabed sediment is that the induced stresses are transient, each peak lasting from only a fraction of a second to a few seconds. The response of the sediment to this type of loading is different from the response under a sustained load. One possible response for a granular material is that the sediment progressively loses most of its strength or liquefies and flows under its own weight resulting in large permanent deformations. Another alternative is that permanent soil deformations occur only during the short period of time that an induced stress peak exceeds a threshold strength. In this case, movement of the seabed caused by the several peaks of stress during a storm or earthquake may not be serious, provided that large deformations have not occurred and the soil strength after the cyclic loading is still adequate to resist the applied static gravity loads. The response of clays and granular material to cyclic loading are very different and will be treated separately in the following sections.

7.3.1 Theoretical Background

The behavior of saturated sediments under cyclic loading is very dependent on whether the effective stress between particles increases or decreases. The effective stress in turn is dependent on whether excess pore water pressure decreases or increases during the cyclic loading. This in turn depends upon whether the void spaces between particles decrease or increase. In order for any change in void space dimension to occur requires displacement of particles. The displacement of particles can result in either a reduction of void space (i.e., contractive soil) or an expansion of void space (i.e., dilative soil). A contractive soil when loaded in compression from its initial state to failure exhibits either a continuing increase in positive pore pressure if undrained or a continuing decrease in void ratio if drained. A typical undrained compression test on a contractive soil is shown schematically as curve I in Figure 7.6 for an undrained condition. For the undrained case point a is the initial state of the soil. A stress path as a compression load is applied is shown leading to point b at the peak shear stress. The test is then continued to failure at point c on the steady-state (SS) line. The SS line represents a state of failure in which the soil deforms continuously at constant volume and constant shear stress. The SS line corresponds to both the Casagrande's constant void ratio (CVR) line for sands and the critical state line (CSL) for monotonic loading used by other authors. The concept was extended to monotonic loading of silts and clays by Poulos (1981). In contrast, the cyclic limit state (CLS) line as shown in Figure 7.6 has been redefined to represent a failure state induced by cyclic loading

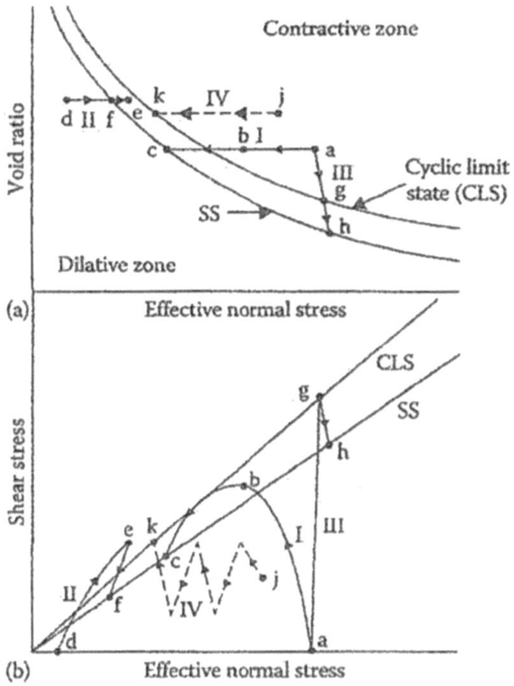

FIGURE 7.6 Perspectives of the state space with several test paths: (a) void ratio versus effective normal stress and (b) shear stress versus effective normal stress (After Sangrey et al., 1978). Reprinted with permission of ASCE.

(France and Sangrey, 1977; Sangrey et al., 1978). The CLS reflects a reference state for failure induced by cyclic loading and constitutes the upper bound of non-failure behavior for very large number of cycles. The SS and CLS line are both independent of stress history.

Compared to the contractive soil a dilative soil prior to failure during a compression test exhibits a decrease in pore pressure or increase in void space depending upon whether the specimen is undrained or drained. In Figure 7.6 curve II is a typical example of an undrained compression test on dilative soil. The test begins at an initial stress represented by point d. Loading is applied along the stress path to the peak shear stress at point e. Further loading results in a stress path to point f on the SS line. Curve III represents the stress path for a typical drained contractive soil. For saturated contractive soil under undrained cyclic loading the stress path is presented in curve IV. The initial anisotropic stress state on the soil is represented as point j. Each successive loading cycle then leads to an further increase in positive pore pressure. This continues until failure occurs when the stress path reaches point k on the CLS line. From point k the soil will continue to deform whenever the shear stress level exceeds the CLS.

The schematic drawing in Figure 7.6 indicates that the SS line and the CLS are different. In reality, this depends on whether the material is a clay or a sand. For a clay, the SS is associated with the residual friction angle while the CLS is approximately equal to the remolded friction angle. The remolded friction angle will be significantly higher than the residual friction angle. As a result, the CLS will be the remolded strength state for clays (Sangrey et al., 1978). In contrast, for a sand-like material the CLS corresponds to the maximum friction angle state. Therefore, the CLS and the SS should be approximately the same (Sangrey et al., 1978).The normalized stress–strain and pore pressure–strain curves for undrained cyclic loading triaxial tests on specimens of both sand and clay is presented in Figure 7.7. A review of Figure 7.7 shows that cyclic loading on undrained saturated contractive soils results in increasing excess pore pressure and axial strain. A comparison of the state paths of contractive specimens of sand and clay when subjected to cyclic loading is presented in Figure 7.9 A review of Figure 7.9a and 7.9b for clay and sand, respectively, shows that the CLS line is considerably flatter for the latter material. This flatness reflects the smaller range of void ratio variation available for sands compared to clays. Therefore the horizontal distance between the initial loading point j and the failure point k on the CLS line is much larger for sands. This length is directly related to the amount of excess pore water pressure change that occurs. The excess pore pressure leads to an effective stress at the CLS. The stress paths for both of these conditions are shown in Figures 7.9c and 7.9d . A review shows that the principal stress difference is much larger for the clay as compared to the sand because of the difference amount of excess pore pressures that was developed

7.3.2 Threshold Strain Concept

Low strain soil behavior can be illustrated by considering the soil as an assemblage of discrete elastic particles. A cubically packed array of spheres loaded along one of the packing axes is shown in Figure 7.8. A packing axis is the preferred orientation of the majority of the particles.

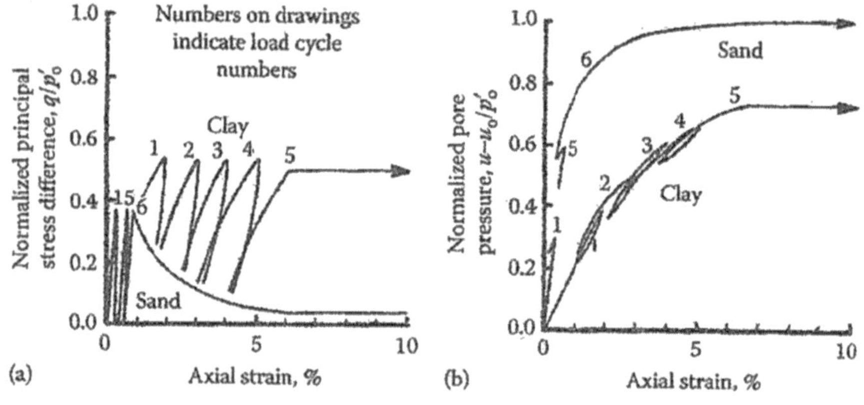

FIGURE 7.7 Cyclic loading of contractive specimens of both sand (a) and clay (b) leads to failure when pore pressures increase sufficiently. Numbers indicate cycle number (After Sangrey et al., 1978). Reprinted with permission of ASCE.

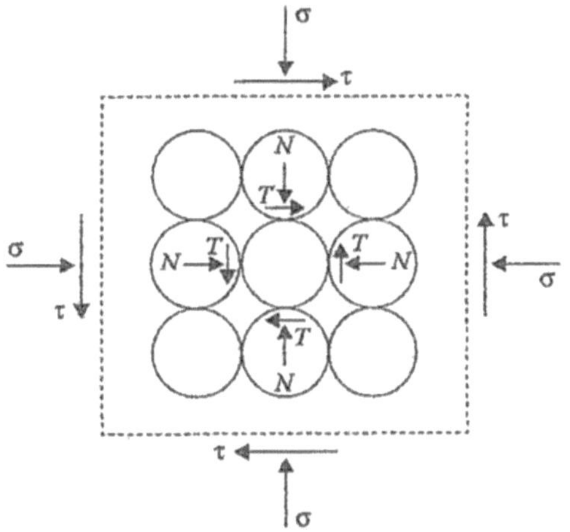

FIGURE 7.8 Cubically packed assemblage of spheres subjected to normal stress, σ, and shear stress, t, which produce inter-particle contact forces N and T (After Dobry et al., 1982). *Prediction of Porewater Pressure Buildup and Liquefaction of Sands during Earthquakes by the Cyclic Strain Method*, NBS Building Science Series 138, US Department of Commerce, National Bureau of Standards, 1982.

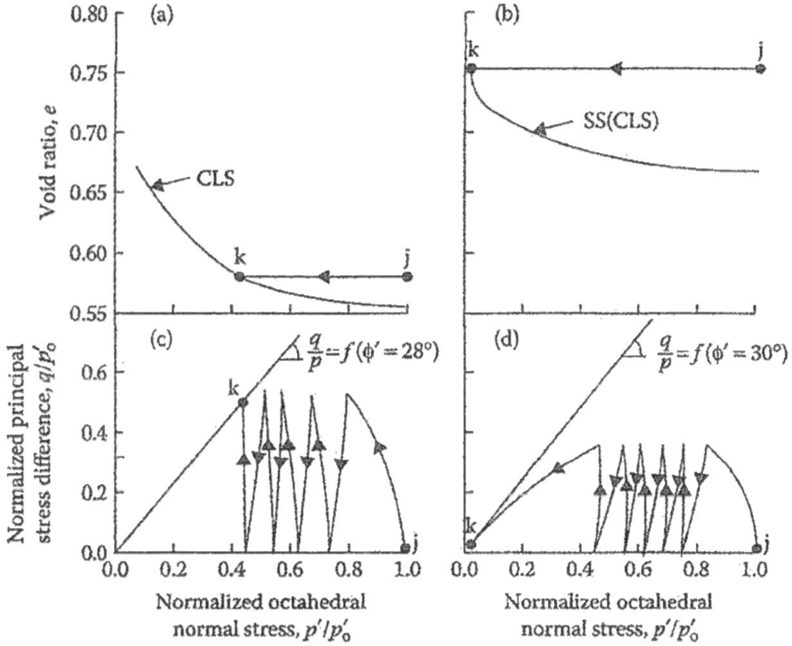

FIGURE 7.9 A comparison of the state paths of contractive specimens of sand and clay when subject to cyclic loading illustrates the difference in pore pressure response (After Sangrey et al., 1978). Reprinted with permission of ASCE.

When a tangential force, T is applied, elastic distortion causes the centers of the spheres to be displaced perpendicular to their axis (Mindlin and Deresiewicz, 1953; Dobry et al., 1982) by an amount given in the equation below:

$$\delta_T = \left[1-\left(1-\frac{T}{fN}\right)^{0.67}\right]\left[\frac{3fN}{4E}(2-v)(1+v)\left[\frac{3\left(1-v^2\right)NR}{4E}\right]^{-0.33}\right] \tag{7.5}$$

where f is the coefficient of friction between the spheres.

When T becomes equal to fN, gross sliding of the particle contacts occurs. This gross sliding is required for permanent particles reorientation as shown in Figure 7.10. This reorientation of particles results in either volume changes (drained conditions) or excess pore pressures (undrained conditions). If particle reorientation does not occur then neither volume change nor excess pore pressure will occur.

The shear strain corresponding to the initiation of gross sliding is called the volumetric threshold shear strain. This relationship is given below assuming $T = fN$.

$$\gamma_{tv} = 2.08\frac{(2-v)(1+v)f}{\left(1-v^2\right)^{0.5}E^{0.67}}\sigma^{0.67} \tag{7.6}$$

Kramer (1996) showed that if the properties of quartz ($E = 7.6 \times 106$ kPa, $v = 0.31$, $f = 0.50$) are substituted into the above equation, the threshold shearing strain is given by the following:

$$\gamma_{tv}(\%) = 1.76\times10^{-7}\sigma^{0.67} \tag{7.7}$$

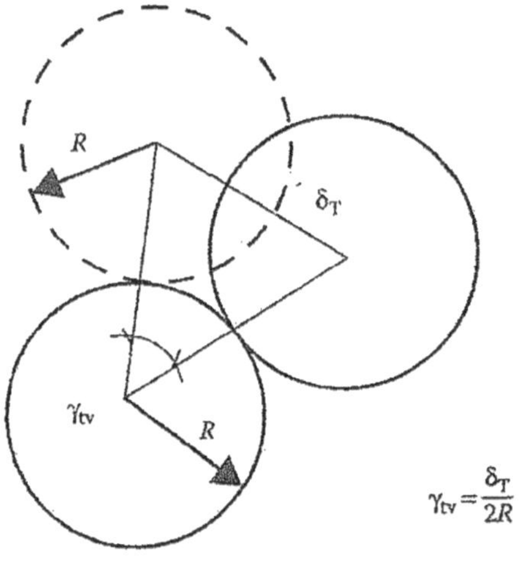

FIGURE 7.10 Particle reorientation.

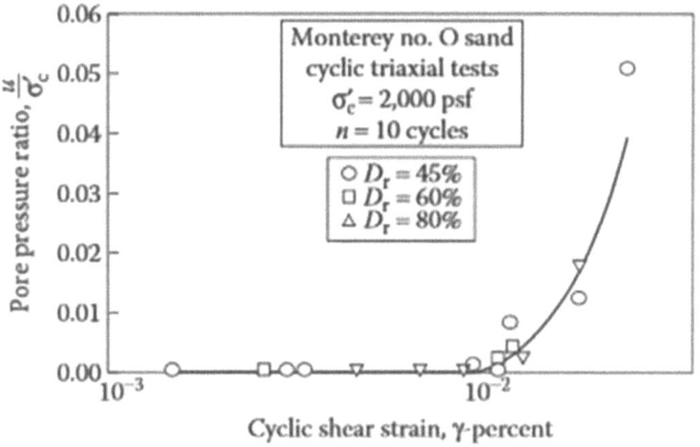

FIGURE 7.11 Pore pressure as a function of cyclic shear strain illustrating a threshold strain of about 0.01%, below which no excess pore pressure are developed (After Dobry et al., 1982).

For confining pressures of between 25 and 200 kPa the equation would predict a threshold shear strain between 0.01% and 0.04%. The existence of a threshold shear strain very close to that predicted by the equation has been observed experimentally for sands under both drained (Drnevich and Richart, 1970; Youd, 1972; Pyke, 1973) and undrained (Park and Silver, 1975; Dobry and Ladd, 1980; Dobry et al., 1982) loading conditions (Figure 7.11). Experiments have also shown that the threshold shear strain increases with plasticity index (PI). The volumetric threshold shear strain of a clay with PI = 50 is approximately an order of magnitude greater than that of a sand with PI = 0 (Vucetric, 1994). Experimental evidence from resonant column (RC) tests also shows that soils exhibit linear elastic behavior below a linear cyclic threshold shear strain, γ_{tl}, that is approximately 30 times smaller.

7.3.3 Clay Minerals under Cyclic Loading

The behavior of clay soil to cyclic loading will typically involve a possible decrease in undrained shear strength, generation, and dissipation of excess pore pressures, modification of stiffness and damping, and accumulation of permanent strains. The behavior of clay materials under cyclic loading is illustrated by two tests conducted by Sangrey et al. (1969). Sangrey conducted cyclic triaxial (CTX) tests utilizing very slow axial compression and rebound loading on clay soil. Results show that after the first loading cycle a net non-recoverable deformation and a net increase in pore water pressure remained. Upon the second load cycle a further increase in residual pore water pressure and axial strain occurred. Subsequent cycles of load and unload application repeated this pattern until on the 10th cycle the sample was unable to carry the load and the rate of non-recoverable deformation increased.

In addition, maximum pore water pressure was reached before the maximum stress difference $(\sigma_1 - \sigma_3)$ except for cycle number 1.

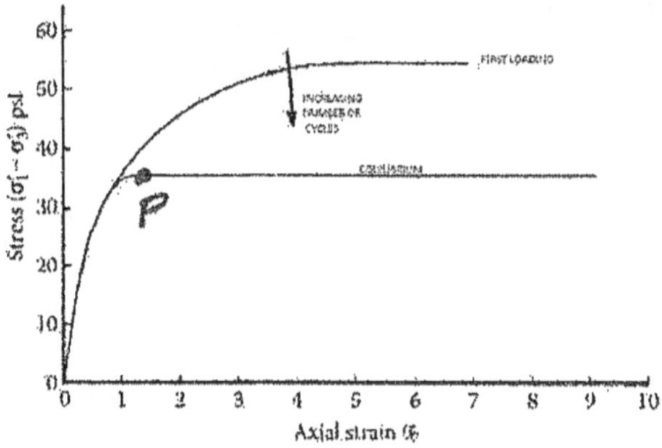

FIGURE 7.12 General stress-strain behavior of clays under cyclic loading (Adapted from Sangrey et al., 1969).

The effective stress path for the above loading is also shown in Figure 7.12. The space presented is a plane representing axial symmetry in which axial stress, σ_1', is shown in the vertical.

Results of a smaller cyclic loading on a clay soil similar to that described previously was discussed. The deformation and pore water pressure changes occurring during the first cycle of loading shows that unlike the previous example the behavior upon unloading results in a decrease in the resulted in increasing permanent deformation and net pore water pressure until a maximum value was reached after six cycles. Additional cycles of load application resulted in no further increase. This shows that both the stress–strain and pore water pressure–strain curves formed closed hysteresis loops. Sangrey named this state as non-failure equilibrium. The general stress–strain behavior of clays under cyclic loading is presented in Figure 7.12. A review of this figure shows that up to a point P the curve representing the locus of equilibrium coincides with non-equilibrium. Beyond the peak point P the non-equilibrium response requires a higher stress difference to be applied to achieve a given axial strain in the specified number of cycles. Therefore at point P equilibrium is no longer possible and the cumulative strains tend toward infinity. The strength under cyclic loading is therefore less than exhibited under static loading.

A summary of the effect of cyclic loading on the behavior of a fine-grained material can be schematically visualized as shown in Figure 7.13. In this figure, three loading conditions (A, B, and C) are presented. Loading condition A corresponds to a statically applied load. In contrast, loading condition B is a cyclically applied load, which has reached an equilibrium resulting in a situation where neither the pore water pressure ($\delta\mu$) nor strain (ε) is increasing. Loading case B would model a seabed sediment that has reached an equilibrium state while undergoing a wave loading. Loading case C represents a cyclically applied load in which the pore water pressure is increasing and consequently the strain. Loading case C would model a seabed sediment that is proceeding toward a failure situation while undergoing a wave

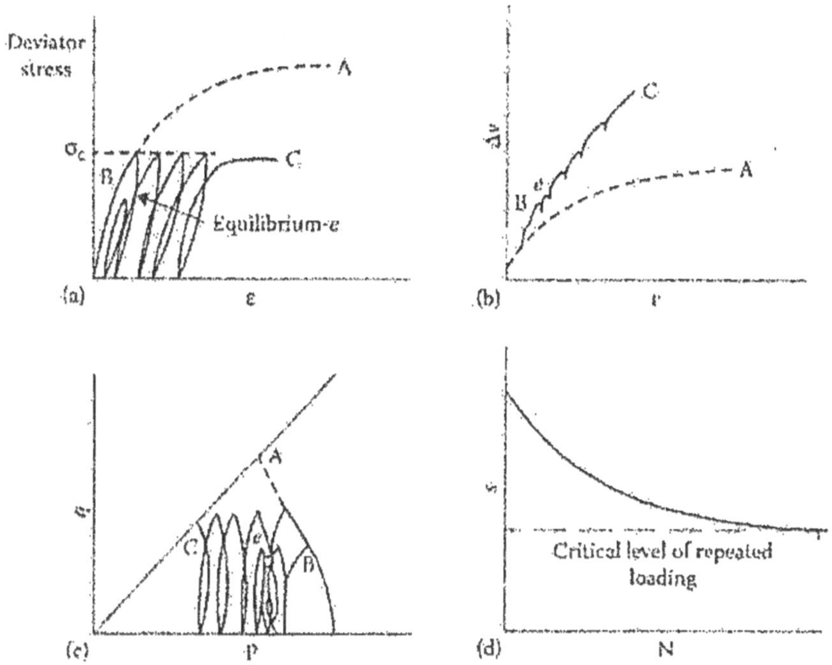

FIGURE 7.13 (a–d) Summary of the response of saturated clay soils related to repeated loading (After Sangrey, 1977). Reprinted with permission of Taylor & Francis Group.

loading. The position of stress path in relation to failure envelope SSL (SS of deformation) is reached only at large strains, thus obscuring the effects of soil fabric, stress and strain history, and loading conditions.

7.3.4 Sands under Cyclic Loading

The response of sands to cyclic loading will typically consider liquefaction, magnitude of excess pore water pressure generation, changes in stiffness and damping, cyclic strains and both the resulting cyclic displacements and permanent displacements. Sands accumulate positive pore pressure during cyclic loading. These pore pressures may be large enough to cause stress path to travel along the failure line. Unlike contractive soils dilative soils do not experience continuing deformation or large strength reduction.

7.4 SOIL PARAMETER MODELING

The effects of cyclic loading are extremely dependent upon the soil and nature of cyclic deformation as shown in the previous section. The potential for pore water pressure generation in clays, silts, or sands governs their behavior under cyclic loading. Two analytical approaches have been developed to model this behavior.

1. Total stress–strain approaches in which the cyclic strain or stress conditions of interest are simulated on laboratory specimens and behavior inferred directly
2. Effective stress–strain approaches in which one attempts to determine analytically the pore water pressures generated and their influence on the soil behavior. Laboratory tests on the sediments are utilized at a more fundamental level to determine pore pressure generation characteristics

The effective stress approaches are more intuitively satisfactory from a fundamental standpoint. However, in present offshore practice it is not common that enough information is known about either the in-situ stress condition or the stress–strain pore water pressure characteristics of the material under cyclic loading conditions to enable effective implementation of this approach.

The total stress–strain approaches for cyclic loading problems have largely evolved from work on earthquake response-related problems. In general, two methodologies have evolved using the total stress–strain approach. The first method involves the development of parameters for input into a stress–strain matrix such as employed in a typical finite-element simulation.

In the second approach an attempt is made to model the in-situ condition in the test apparatus and the resulting deformations are observed. A discussion of the two methodologies as applied to prediction of the deformation of the seabed under cyclic loading is presented by Chaney (1984) and cyclic degradation models for soft and marine clays are given by Chaney and Fang (1984). These two approaches along with their method of application to analytical technologies are presented in Figure 7.14 (Byrne, pers. comm.).

In utilizing the first total stress–strain approach, two predominant changes in the physical behavior of the soft clay are observed. These changes are (1) a decrease in the stiffness (modulus) and (2) an increase in the amount of energy being consumed per cycle of loading (damping). This behavior is shown graphically in Figure 7.15 for clay material from the Gulf of Mexico as it undergoes a constant stress CTX loading test. In this plot, the stiffness or modulus using the RC at low shearing strain levels ($<5 \times 10^{-2}\%$) and a CTX apparatus at higher levels of strain. The modulus from a CTX test can be determined from the slope of a straight line drawn through the two end points of a single hysteresis loop. The modulus is observed to decrease with increasing cycles of loading. In addition, on each cycle of loading the hysteresis loop becomes larger and larger indicating that the amount of energy being consumed is increasing.

The relationship between the shear modulus ratio (G/G_{max}) versus the log of shear strain ratio (γ/γ_r) for various soil types including clay, sand, and silt is shown in Figure 7.16. G_{max} is the maximum shear modulus that occurs at an approximate shearing strain level of $10^{-4}\%$. The ranges of G/G_{max} for marine deposits shown in Figure 7.16 indicate that a large number of marine deposits are silt-like materials and their behavior is similar to sand rather than clay (Fang et al., 1981; Fang and Chaney, 1986). A hyperbolic curve was fitted to the normalized shear modulus versus shear strain variation data of Gulf of Mexico clay, where the agreement was shown for the

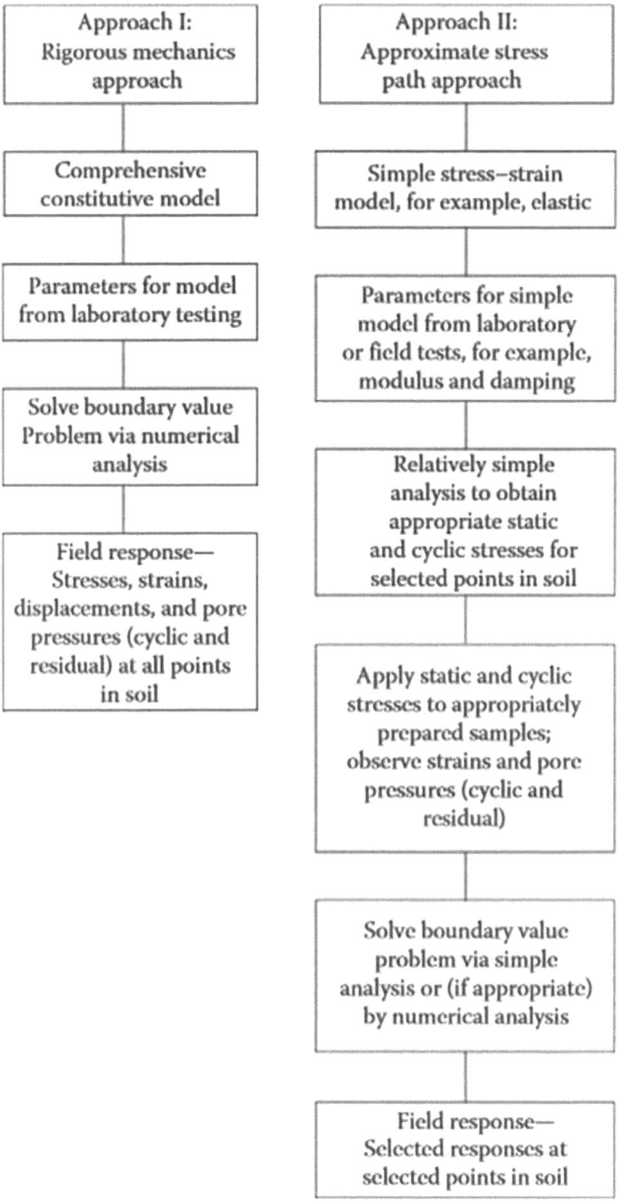

FIGURE 7.14 Alternative approaches to the analysis of cyclic response (Data from Byrne, P.M., Pers. comm. to Prof. H.G. Poulos, as presented in Poulos, 1988).

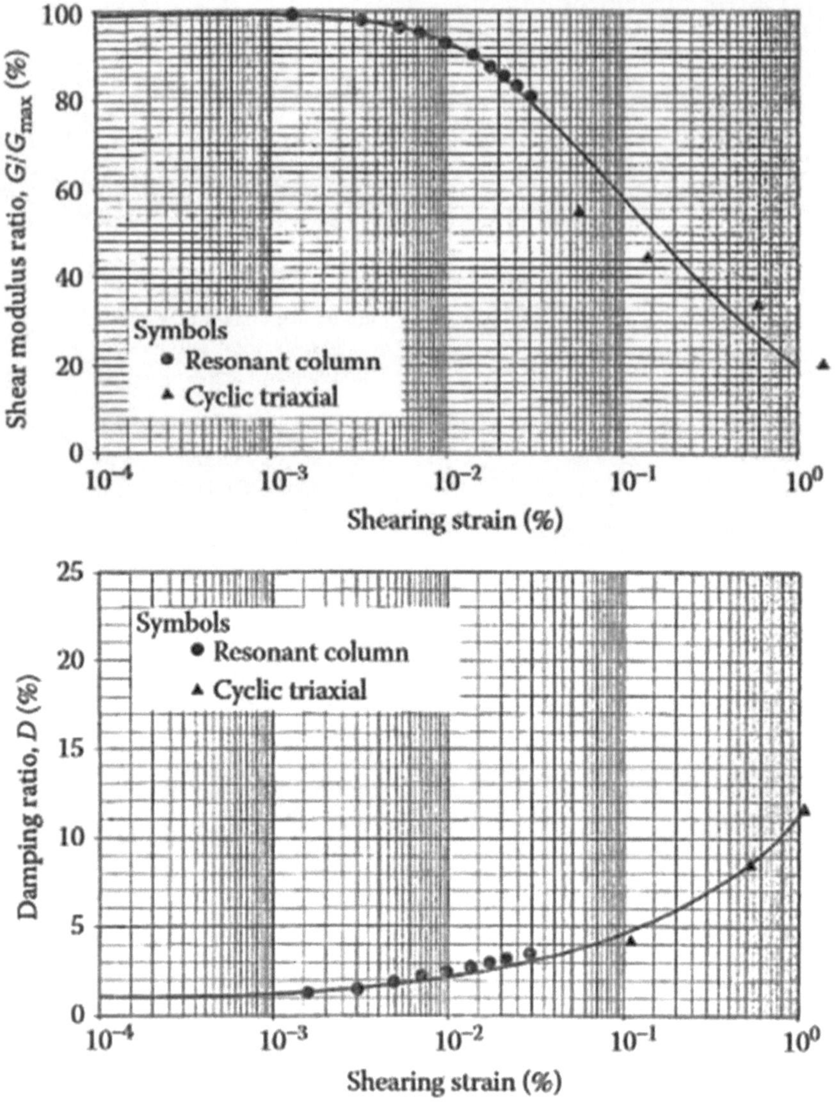

FIGURE 7.15 Results of cyclic triaxial test on Gulf of Mexico clay.

lower ranges of strain amplitude normalized by yield strain (Pamukcu et al., 1983). A comparison of G_{max} versus depth for variety of marine sites has been presented by Anderson et al. (1983) and later expanded by Chaney (2013) in Figure 7.17. A review indicates that maximum shear modulus increases as a function of depth and varies widely from location to location.

Cyclic tests using the second approach may be presented either in the form of (1) a normalized cyclic strength versus the number of cycles to a specified constant

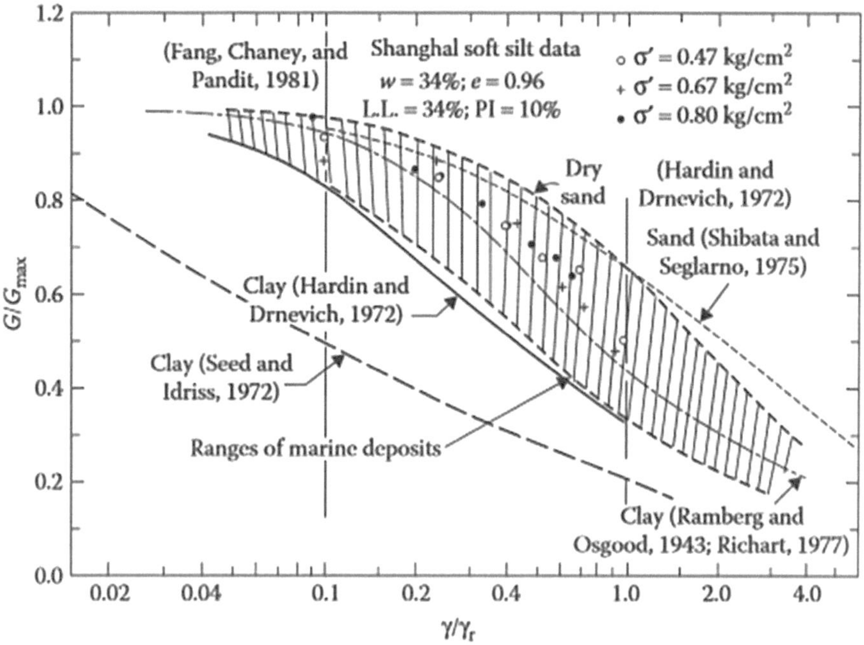

FIGURE 7.16 Comparison of dynamic shear modulus ratio (G/G_{max}) versus strain ratio (γ/γ_r) for various soil types and ranges of marine deposits (Data from Fang, H.Y. et al., 1981).

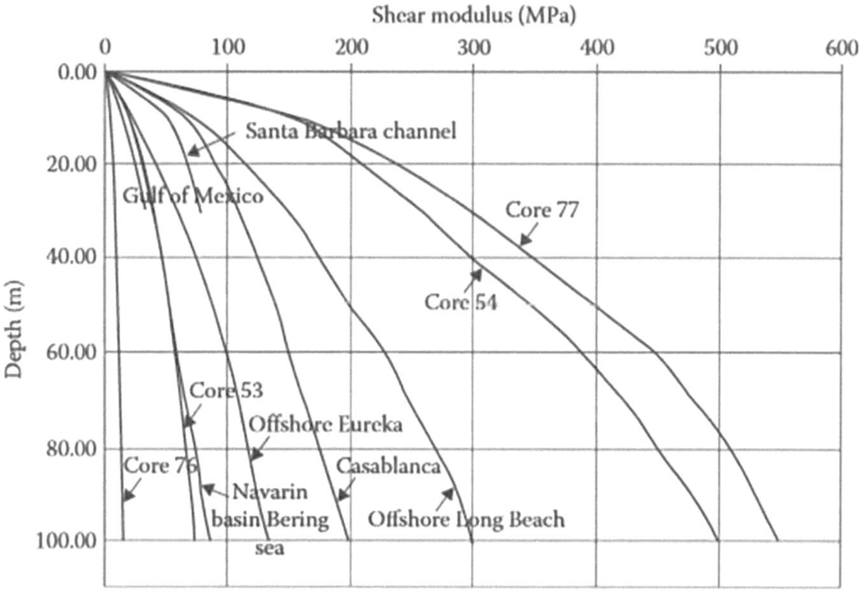

FIGURE 7.17 Maximum shear modulus, G_{max} versus depth profiles (From Chaney, 2013). Reprinted with permission. Copyright ASTM.

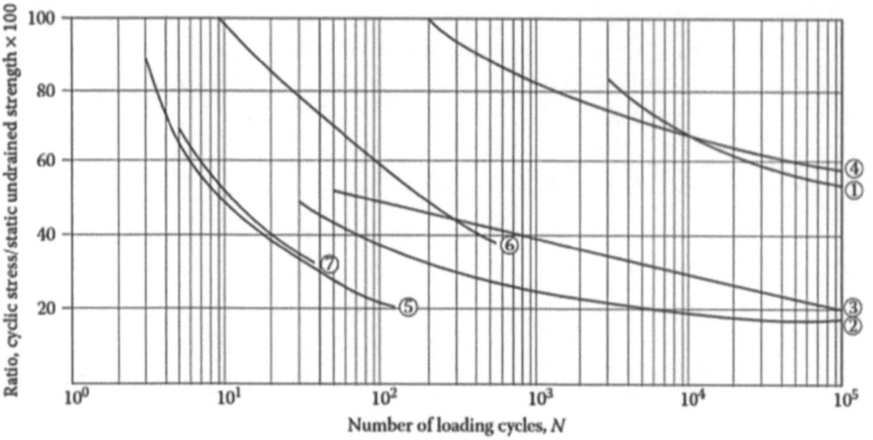

No.	Material	Apparatus	Reference
1	Pacific pelagic clay	Triaxial	Houston and Herrman (1980)
2	Pacific pelagic clay	Triaxial	Houston and Herrman (1980)
3	Pacific hemi-pelagic	Triaxial	Houston and Herrman (1980)
4	San francisco bay mud	Triaxial	Houston and Herrman (1980)
5	Vicksburg silty clay	Triaxial	Seed and Chan (1966)
6	Ekofisk sandy clay	Simple shear	NGI
7	Anchorage clay	Simple shear	Thiers and Seed

FIGURE 7.18 Compilation summary of cyclic strength of clays, pure stress reversal (From Chaney and Fang, 1986). Reprinted with permission. Copyright ASTM.

strain as shown in Figure 7.18 or (2) a normalized diagram as illustrated by Figure 7.19. A review of Figure 7.18 shows that there is a wide range of responses of marine clays to cyclic loading.

An example is to compare the behavior of a Pacific pelagic clay (No. 1) at a stress ratio of 60% with and Ekofisk sandy clay (No. 6). The number of cycles of loading N at this stress ratio are 30,000 to 100, respectively. In terms of the marine environment, this indicates that the Ekofisk site is more susceptible to the effects of cyclic loading than a corresponding site composed of Pacific pelagic clay. This type of cyclic loading analysis has been presented by a number of authors for a variety of soil types. A small selection of authors who have contributed are the following: Seed and Chan (1966), Herrmann and Houston (1976, 1978), and Anderson et al. (1980). Figure 7.20 shows a boundary line between a stable and an unstable region for a given number of applied cycles. Outside the boundary, the imposed stress conditions lead to a progressive accumulation of permanent (average) strains or a progressive increase in the cyclic strain amplitude before the given number of stress cycles is reached. Inside the boundary, the situation is stabilized after the first few cycles. In the following sections the behavior of clays and silts will be discussed first followed by sands.

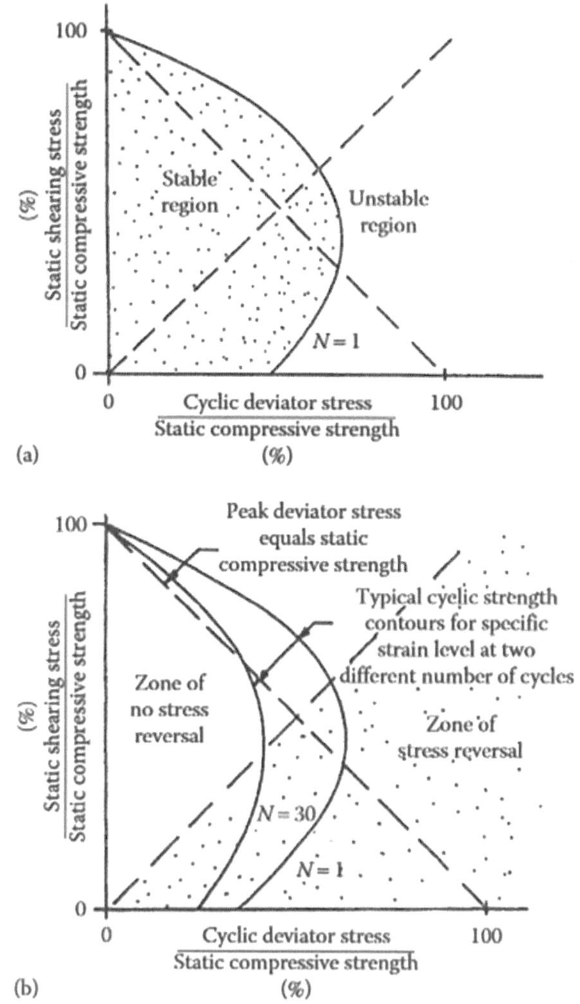

FIGURE 7.19 (a, b) Schematic of cyclic strength contour methodology (After Chaney and Fang, 1986). Reprinted with permission. Copyright ASTM.

7.5 BEHAVIOR OF CLAYS AND SILTS

Clays and silts subjected to wave and earthquake loading may undergo the following conditions: (1) the deterioration of undrained shear strength, (2) the degradation of stiffness, (3) the generation of excess pore pressures and their subsequent dissipation, and (4) the accumulation of permanent strains. Some of the factors that affect cyclic strength of clays and silts are cyclic stress level and its frequency, number of cycles of loading, initial shear stress, stiffness, effective confining pressure, and overconsolidation ratio (OCR).

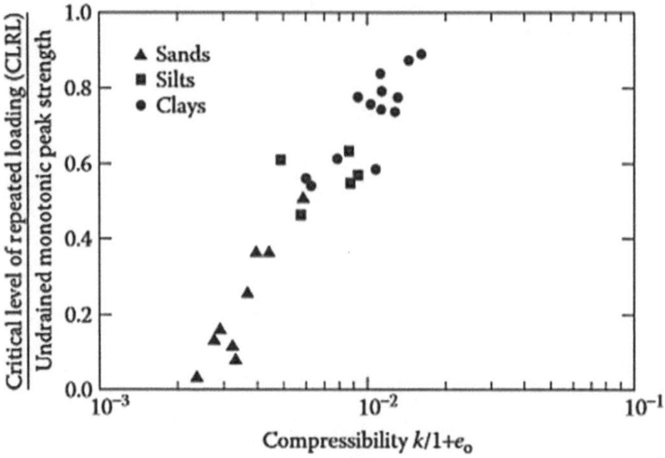

FIGURE 7.20 Critical level of repeated loading (CLRL) from undrained tests on contractive soils (After Sangrey et al., 1978). Reprinted with permission of ASCE.

Sections 7.5.1 through 7.5.3 discuss (1) strength determination, (2) pore pressure build-up, and (3) reduction and degradation of stiffness.

7.5.1 Strength Determination

During cyclic loading, excess pore pressures can build up, causing large cyclic strains. These strains may become so large that the soil would be considered to have failed. The cyclic stress to cause failure can be related to the number of cycles of loading, which often shows a fatigue-type behavior (Lee and Focht, 1976). It has also been shown that when clayey soils are loaded with initial shear stress or a fraction of their static failure stress, this may increase their cyclic shear strength significantly (Ishihara and Yasuda, 1980). The effect of strength deterioration due to cyclic loading is more pronounced when the stress reversal causes the soil to experience negative stresses at each cycle of loading. In addition, strength is also correlated with PI in which cyclic strength at zero initial shear stress normalized by static strength increases with PI.

Sangrey et al. (1978) suggested that contractive clays, silts, and sands behave similarly under repeated loading. Figure 7.20 shows the correlation between the recompression index and normalized critical level of repeated loading (CLRL) for clays, silts, and sands, where e_0 is the initial void ratio. CLRL is the cyclic stress level at which soil passes into failure from an equilibrium state. This correlation indicates that larger levels of cyclic stress are sustained by clays owing to their larger value of recompression index. Meimon and Hicher (1980) correlated the post-cyclic undrained shear strength of two laboratory-prepared clays (with OCR = 1 and OCR = 4), with maximum permanent axial strain. They found that if the maximum permanent strain did not exceed 5%, the reduction in undrained shear strength was lower than 8%. With increasing strain level, the reduction could reach up to 40%. Experimental results also showed that the reduction did not necessarily depend upon OCR. The effect of loading rate has also been shown to influence cyclic behavior of clays. The effect of loading rate on pre- and post-cyclic undrained shear strength can be approximated by the following expression (Poulos, 1988):

$$\frac{S_u}{S_{ur}} = 1 + F_r \log_{10}\left(\frac{r}{r_r}\right) \tag{7.8}$$

where

S_u is the undrained strength
S_{ur} is the reference value of Su for reference loading rate r_r
r is the actual loading rate
F_r is the rate factor (typically in the range of 0.05–0.2)

This expression indicates that the undrained strength increases with increasing loading rate. An analytical model proposed by Prevost (1977) looks at the effect of loading rate on the cyclic behavior of a clay soil. The model parameters were determined through slow monotonic and rapid cyclic simple shear tests of Drammen clay. Also shown in this figure are typical hysteresis loops obtained at constant strain amplitudes. The rapid cyclic stress–strain curve lies above the static curve, which indicates increase in stiffness and strength with increasing loading rate. The hysteresis loops develop an S shape, which becomes more marked as the number of cycles of loading increases. It was reported that the gradient of the hysteresis loops at the peak shear stress remained constant and was approximately equal to the gradient of the static curve at the corresponding strain. The effect of number of cycles of loading and frequency of loading on cyclic stress ratio (= t/c; where t = cyclic stress; c = undrained shear strength) were also studied by Procter and Khaffaf (1984). Figure 7.23 illustrate these effects. In Figure 7.20 the data indicate that, irrespective of strain amplitude, there is a minimum stress ratio achieved after a certain number of cycles of loading. A similar phenomenon was observed with shear modulus of soft marine clays, where the data indicated a minimum value of shear modulus after a critical value of number of cycles of loading (Pamukcu et al., 1983). In both cases this value of N (= number of cycles of loading) was around 10,000 and the data were obtained from strain or displacement-controlled cyclic tests. Figure 7.21 also shows effect of loading frequency on the cyclic stress ratio, where the ratio appears to be constant for frequencies below 0.1 Hz and decreases above this value.

7.5.2 Pore Pressure Build-Up

Generation of excess pore water pressures under cyclic loading has been shown to cause marked reduction in undrained strength and stiffness of clay soils. Theoretical and empirical expressions have been developed that relate excess or residual pore pressures with factors such as cyclic stress and strain level, number of cycles of loading, and OCR. An empirical expression has been developed by Van Eekelen and

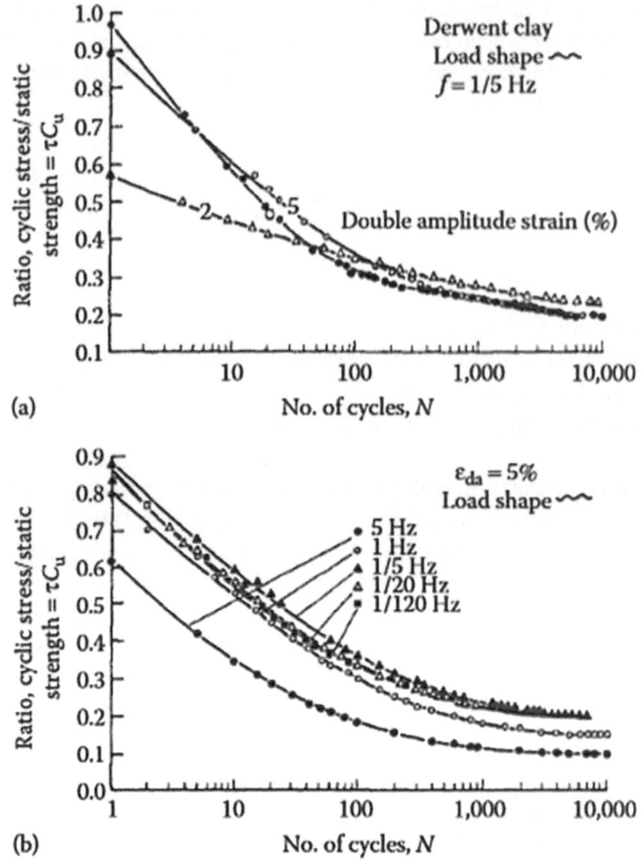

FIGURE 7.21 Variation of cyclic stress ratio with number of cycles of loading, N, in reversed displacement-controlled tests of remolded clay. (a) Constant frequency and (b) constant stress (After Procter and Khaffaf, 1984). Reprinted with permission of ASCE.

Potts (1978) for the rate of generation of excess pore pressures. Another has been developed by Matsui et al. (1980) for the residual pore pressures:

$$\frac{u_r}{\sigma_c} = \beta\left[\log_{10}\left[\frac{\gamma_{c,\max}}{A_1(OCR-1)} + B_1\right]\right] \tag{7.9}$$

where

u_r is the residual pore pressure
σ_c is the effective confining pressure
$\gamma_{c,\max}$ is the single amplitude maximum cyclic shear strain
OCR is the over consolidation ratio
$\beta = 0.45$ (found experimentally) and

PI	A	B
20	$N_G = 0.027\sqrt{PI}$	0.6×10^{-3}
40	1.1×10^{-3}	1.2×10^{-3}
55	2.5×10^{-3}	1.2×10^{-3}

again, found experimentally.

Togrol and Guler (1984) suggested an empirical relation for normally consolidated clay that related the deviatoric stress at failure to excess pore pressure developed during cyclic loading:

$$q_f = 0.63 p_c - 0.39 u_{din} \tag{7.10}$$

where

q_f is the deviatoric stress at failure
p_c is the consolidation pressure
u_{din} is the excess pore water pressure

Using experimental values of maximum excess pore pressure developed and Equation 7.7, they found that the maximum reduction in undrained shear strength of the soil would be in the order of 35% under repeated load application. Singh et al. (1978) indicated that build-up of excess pore pressure during cyclic loading of fine-grained soils produces an effect similar to increasing the soil's OCR in the unloading portion of consolidation. A similar conclusion was drawn by Anderson et al. (1980) where it was shown that the stress path of a statically sheared clay following cyclic loading is similar to that of an over-consolidated clay. This behavior was modeled to predict the undrained cyclic response of slightly over-consolidated clays using results of tests on normally consolidated clays (Azzouz et al., 1989). The model was based on AOCR, apparent over-consolidation hypothesis. The method was shown to provide good estimates of number of cycles to failure and development of excess pore pressures with respect to N. Theoretical expressions derived by Egan and Sangrey (1978) relate residual excess pore pressure (due to plastic strain) and also maximum excess pore pressure developed in fine-grained soils under cyclic loading to critical-state soil parameters.

$$du_r = \left[1 - \exp\left(\frac{-\pi}{\kappa}\right)\right] p_0 \tag{7.11}$$

$$du_{\max} = \left[1 - \exp\left(\frac{-\pi}{\kappa}\right)\left(1 - \frac{M}{3}\right)\right] p_0 \tag{7.12}$$

where

κ is the rebound or recompression index on the natural logarithmic scale (=Cr/2.3)

π is the volume change potential [$=k \ln$ (po/pu)], where pu= mean effective stress at critical state
p_0 is the initial mean effective stress
M is the critical-state soil parameter (slope of CSL in p–q space)

Ansal and Erken (1989) developed an empirical model that relates cyclic stress ratio and pore water pressure ratio for different values of N. Figure 7.22 illustrates this relation. A threshold cyclic stress ratio is defined that is similar to observations reported by Matsui et al. (1980) and Dobry et al. (1982), the latter being in sands and in terms of cyclic strain. The relationships reported by Ansal and Erken were established through a series of cyclic simple shear testing of normally consolidated kaolinite clay. In this model the pore pressure build-up is expressed as

$$u = \left[\left(\frac{\tau}{\tau_f} \right) - (S.R.)_t \right] m \tag{7.13}$$

where
u is the pore pressure buildup
τ/τ_f is the cyclic shear stress ratio, where τ_f = cyclic shear stress at failure
$(S.\,R.)_t$ is the threshold cyclic stress ratio
m is the slope of the pore pressure line [$=Du/D(\tau/\tau_f)$]

The effect of frequency of loading on pore pressure build-up was also investigated. It was found that a decrease in loading rate leads to an increase in the accumulated

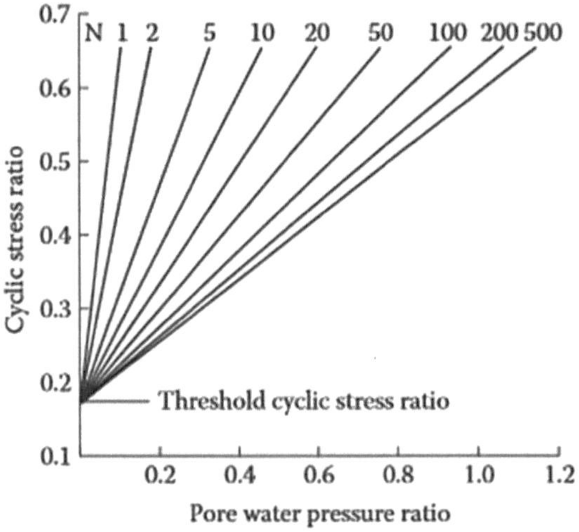

FIGURE 7.22 Cyclic stress ratio-pore pressure relationship for different numbers of cycles (After Ansal and Erken, 1989). Reprinted with permission of ASCE

pore pressures with a number of cycles and the effect of loading rate diminished after the initial cycles of loading. An indirectly related observation was made by Pamukcu and Suhayda (1987) in which the stiffness degradation of soft saturated clays with induced initial pore pressure ratios was more marked in slow monotonic loading than high-frequency dynamic loading for low strain amplitude range. This indicated larger pore pressure build-up in the slow monotonic loading case.

7.5.3 Reduction and Degradation of Stiffness

During cyclic loading, the stress–strain behavior of clays and silts is nonlinear and hysteretic. For San Francisco Bay silty marine clay, consecutive hysteresis loops obtained for the first cycle of dynamic loading obtained at different controlled strain levels illustrates the stiffness reduction as the enlargement of the loops as they deform and tilt with increasing strain amplitude (Figure 7.23) (Idriss et al., 1978). A hysteresis loop consists of three stages during one cycle of loading: the initial loading, unloading, and reloading stages. The initial loading constitutes the backbone curve of the loop. An idealized stress–strain hysteresis loop obtained for a soil specimen subjected to a symmetric cyclic shearing load along a plane free of initial shear stress is given in Figure 7.24. The backbone curve characterizes the nonlinear stress–strain behavior of clays. G_{max} is the maximum shear modulus defined as the slope of the initial tangent to the backbone curve. G_s is the secant modulus equal to the τ_{max}/γ_{max} ratio, where τ_{max}/γ_{max} are the maximum cyclic shear stress and maximum shear strain, respectively. The most widely accepted rule for generating hysteresis loops from backbone curve is the Masing criterion (Masing, 1926). It simply states that the unloading and reloading branches of the loop are on the same backbone curve with both stress and strain scales expanded by a factor of 2 and the origin translated. After the stress reversal, the tangent modulus at the tips of the loop is equal to G_{max}. The backbone curve is expressed in several mathematical formulations, which include

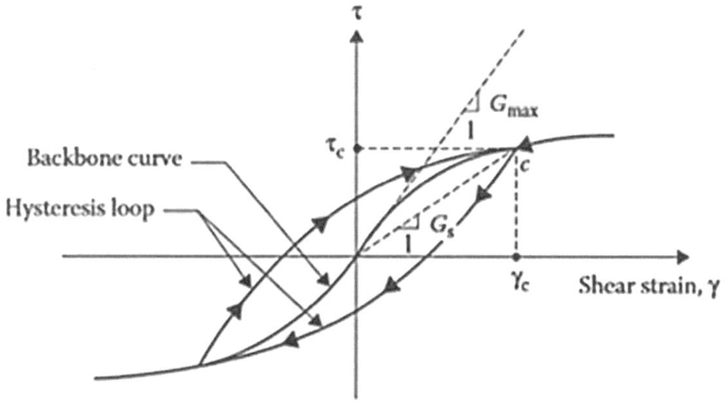

FIGURE 7.23 Typical stress–strain hysteresis loop (After Idriss et al., 1978). Reprinted with permission of ASCE.

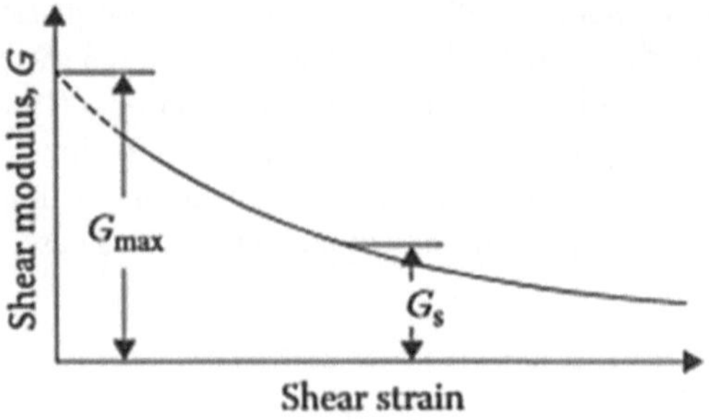

FIGURE 7.24 Idealized shear modulus curve.

bilinear (Thiers and Seed, 1969), multilinear, hyperbolic (Hardin and Drnevich, 1972), and Ramberg and Osgood (1943) formulations.

Reduction of moduli with increasing strain amplitude is a major characteristic displayed by the nonlinear nature of the stress–strain relationship of soils. An idealized shear modulus reduction curve is given in Figure 7.24, whereby extrapolating the curve to zero strain, the maximum shear modulus, G_{max}, can be estimated at the intercept. Hardin and Drnevich (1972) and Hardin (1978) suggested the use of the following form of empirical equation for calculation of laboratory G_{max} for many undisturbed cohesive soils as well as sands:

$$G_{max} = 625 \frac{1}{\left(0.3 - 0.7e^2\right)} \left(\text{OCR}\right)^{\kappa} P_a^{0.5} \sigma_o^{'0.5} \tag{7.14}$$

where

e is the void ratio
OCR is the over-consolidation ratio
σ_o' is the mean principal effective stress
k is the a constant value which depends on PI, as given below
G_{max} is the maximum shear modulus
p_a is the atmospheric air pressure

Note: p_a, G_{max}, and σ_o' all need to be in same units.

PI	*K*
0	0.0
20	0.18
40	0.30
60	0.41
80	0.48
≥100	0.50

This equation, however, is said to produce low G_{max} values for void ratios in excess of 2. Hardin and Drnevich (1972) related G/G_{max} to γ_h, hyperbolic strain, through the following equation.

$$\frac{G}{G_{max}} = \frac{1}{1+\gamma_h} \tag{7.15}$$

where

$$\gamma_h = \left[\frac{\gamma}{\gamma_r}\right]\left(1+a\exp\left(\frac{-b_\gamma}{\gamma_r}\right)\right) \tag{7.16}$$

where
 γ_r is the reference strain
 a and b are the soil constants

The suggested values of a and b are given in Table 7.2. Modulus reduction curves for fine-grained soils of different PI and cyclic shearing strain levels have been presented also by Vucetric and Dobry (1991) as shown in Figure 7.25. A review shows that at any given cyclic shear strain the ratio of G/G_{max} increases with increasing PI and decreases with increasing shearing strain level.

7.5.3.1 Effect of Soil Plasticity on Cyclic Response

In clays, reduction of moduli is generally accompanied by degradation of the backbone curve. Progressive degradation of soil stiffness with increasing number of cycles of loading can be defined as progressive softening of the soil. Degradation is known to be mainly a function of the number of cycles of loading. An example of the effect is shown in Figure 7.26, where the hysteresis loops for a Gulf of Mexico clay at controlled stress are plotted. Degradation effects are formulated using degradation index δ, which is the ratio of the secant modulus in the Nth cycle to the initial secant modulus (Idriss et al., 1978). δ is a function of the number of cycles of load, N, and t defined as follows:

$$\delta = N^{-1} \tag{7.17}$$

TABLE 7.2
Values of A and B for Saturated Cohesive Soils

Application	A	B
Modulus	1 + 0.25 log N	1.3
Damping	$1 + 0.2f^{0.5}$	$0.2f\exp(-\sigma_0') + 2.25\sigma_0' + 0.3\log N$

Source: Hardin and Dmevich (1972). Reprinted with permission of ASCE.

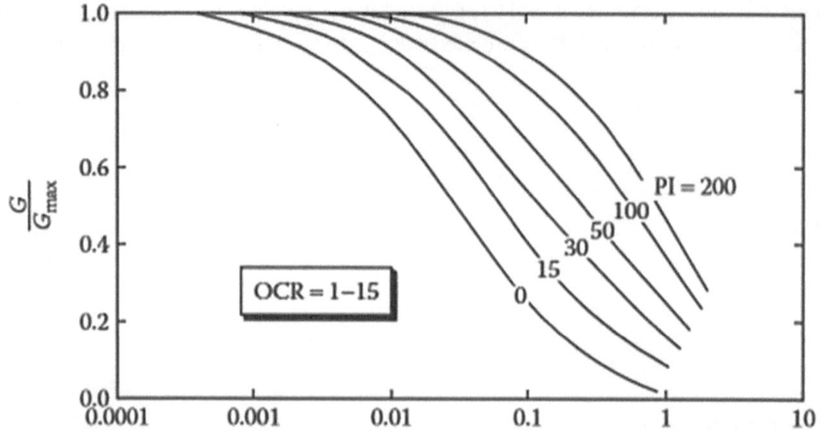

FIGURE 7.25 Modulus reduction curves for fine-grained soils of different plasticity (Data from Vucetric and Dobry, 1991). Reprinted with permission of ASCE.

$$du_{max} = \left[1 - \exp\left(\frac{-\pi}{\kappa}\right)\left(1 - \frac{M}{3}\right)\right] p_0$$

where t is the degradation parameter.

The degradation parameter (t) is defined as the slope of the semi-logarithmic plot of secant moduli ratio versus N. While it is strongly dependent on strain amplitude, it is essentially independent of confining stress, OCR, and water content (Idriss et al., 1978; Singh et al., 1978; Stokoe, 1980; Moses and Rao, 2003). The degradation effect on the backbone curve constructed using the Ramberg-Osgood model is shown in Figure 7.27. A review of Figure 7.27 shows that the backbone curve shifts down and flattens progressively with increasing values of cyclic deviator stress and number of cycles of loading. The variation of the degradation parameter (t) with cyclic stress ratio are shown in Figures 7.28 and 7.29. Moses and Rao (2003) showed for a cemented marine clay that the degradation index (δ) for various levels of effective stress, cyclic stress ratio decreases with increasing number of cycles of load application (Figure 7.28).

Moses and Rao also show that the parameter t can be described by the following relationship for one-dimensional cyclic loading.

$$t = A\left(\frac{q_c}{p_o}\right) + B \tag{7.18}$$

The parameter t is shown to increase with the level of cyclic stress. Test data superimposed on analytical backbone curves, degraded with a number of cycles of loading is shown in Figure 7.30 for Gulf of Mexico clay (Pamukcu and Suhayda, 1984).

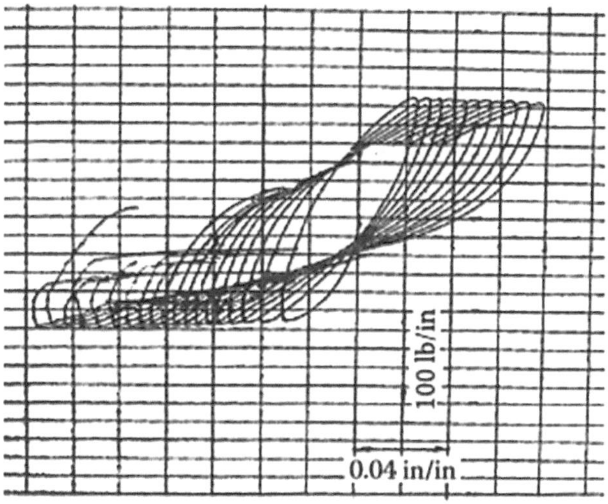

FIGURE 7.26 Hysteresis loops for Gulf of Mexico clay undergoing a CTX test (From Chaney and Fang, 1986). Reprinted with permission of ASTM.

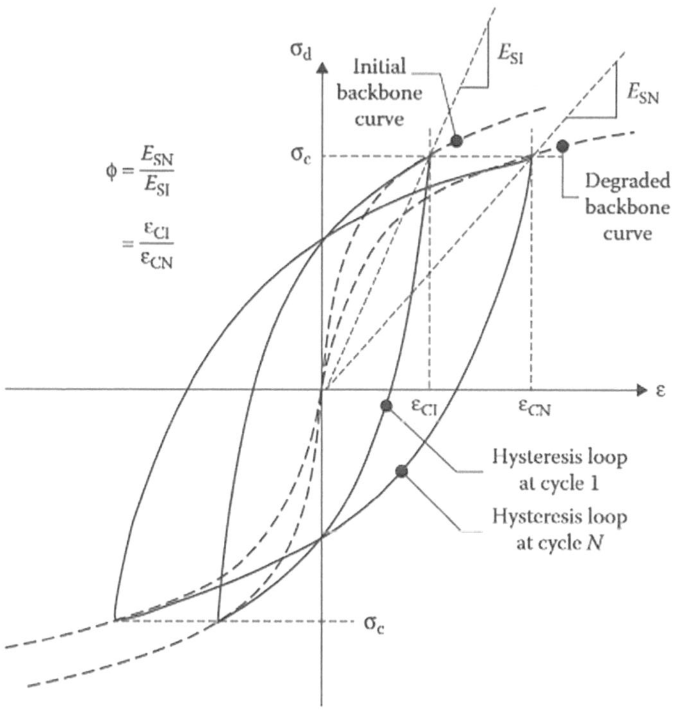

FIGURE 7.27 Schematic illustration of degraded backbone curves (After Moses and Rao, 2003). Reprinted with permission of Taylor & Francis.

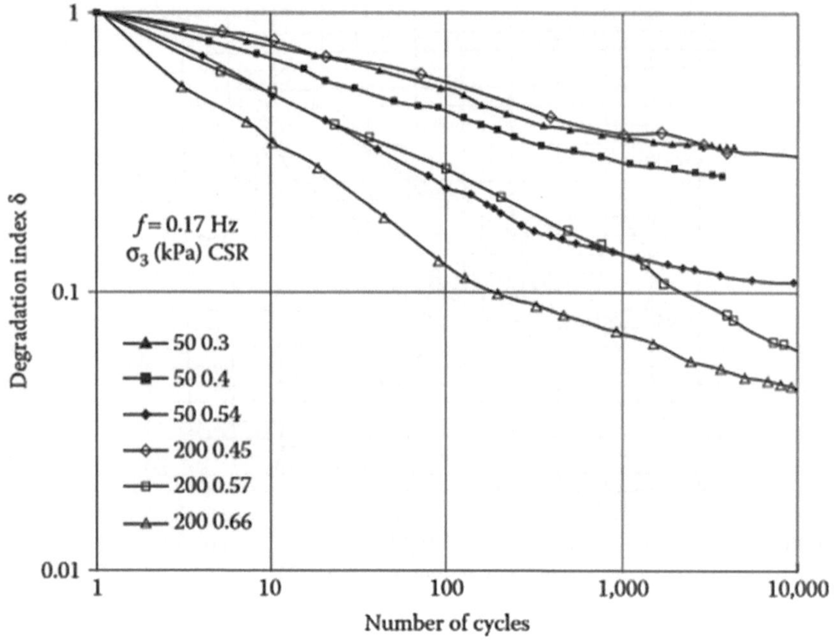

FIGURE 7.28 Variation of degradation index with number of cycles and cyclic stress ratio (CSR) (From Moses and Rao, 2003). Reprinted with permission of Taylor & Francis.

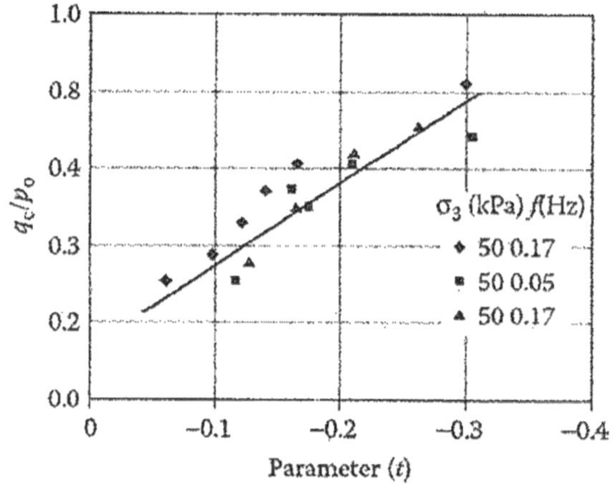

FIGURE 7.29 Variation of t with normalized cyclic stress (From Moses and Rau, 2003). Reprinted with permission of Taylor & Francis.

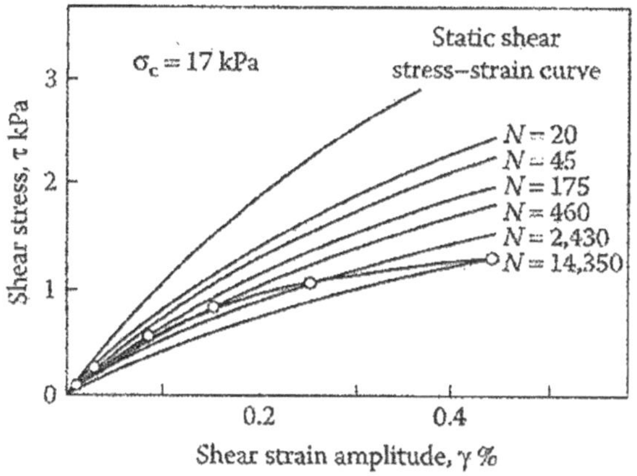

FIGURE 7.30 Degraded backbone curves with actual shear stress–strain data points and the fitted hyperbolic curve (After Pamukcu and Suhayda, 1984). Reprinted with permission of ASTM.

Cyclic degradation of Drammen clay under sustained shear loading is illustrated in Figure 7.32 (Goulois et al., 1985). A review shows that at a constant static imposed shear stress the cyclic shear strength decreases with the number of cycles (Figure 7.31).

The degradation effect diminishes with increasing number of cycles of load application and is shown for different levels of PI in Figure 7.32. Therefore, one can predict a reasonable value of number of cycles at which the material can be assumed to have reached a SS condition with insignificant degradation of stiffness. Degradation and reduction of stiffness is often observed to become significant after a threshold strain amplitude. This threshold of shear strain is generally given between 10%–3% and 10%–2% for clays (Isenhower and Stokoe, 1981; Pamukcu and Suhayda, 1987) as shown in Figure 7.33. Vucetic (1988) has presented information that shows that clays exhibiting static normalized behavior with respect to the vertical consolidation stress, that is, along the lines of the SHANSEP method, also exhibit a similar cyclic normalized behavior. The ratio of the energy dissipated to energy input during one cycle of loading is defined as the damping ratio (D). D is computed based on the area contained within the hysteresis loop, and the equivalent secant modulus. Systems that satisfy the Masing criterion behave as though they have an equivalent viscous damping ratio independent of the frequency of vibration at a given strain amplitude. The effect of various environmental and loading conditions on the damping ratio is presented in Table 7.3. Damping of soils under cyclic loading, like moduli, is strongly dependent on strain amplitude. The tilting of the hysteresis loops in Figure 7.26 is accompanied by the enlargement of the area they enclose as the strain amplitude increases. Since damping, by definition, is the ratio of the area enclosed by the loop to the area under secant modulus line, it can be observed that damping increases with increasing strain (Figure 7.34). Once again, damping ratio stays independent of strain

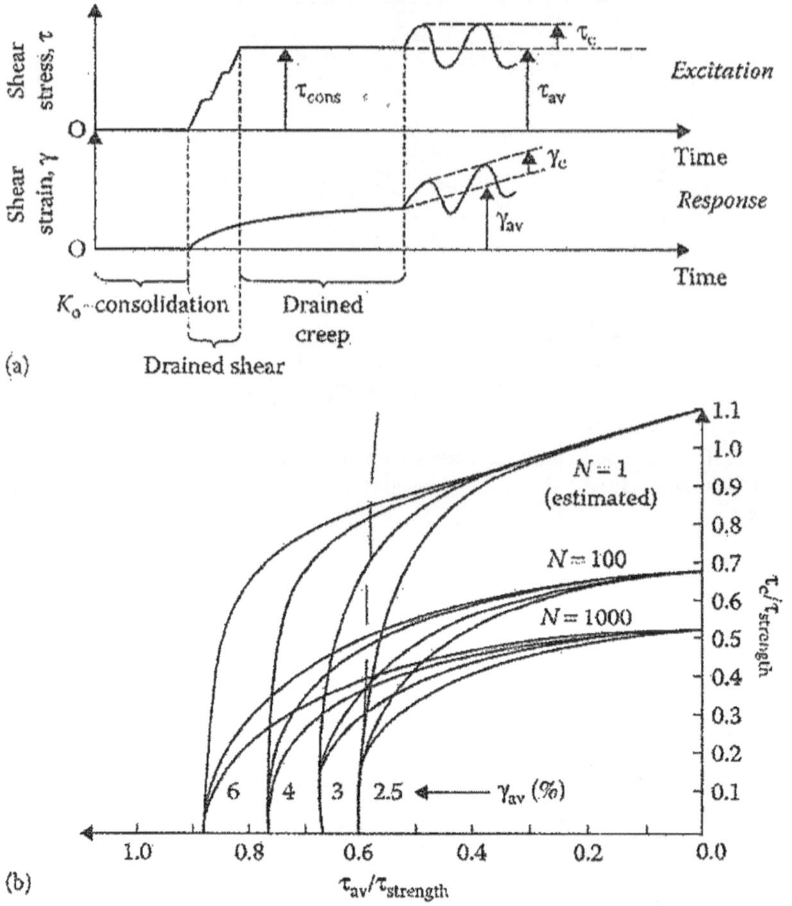

FIGURE 7.31 Cyclic degradation of drammen clay. (a) Shear loading and resulting shear strain, (b) average shear strain network of curves for N = 1100 and 1000 cycles of loading (After Goulois et al. 1985). Reprinted with permission of ASTM.

amplitude up to a threshold value of the cyclic shearing strain, which is between 10–3% and 10–2% for most clays. This value of the damping ratio is called the minimum damping, D_{min}.

7.5.3.2 Factors That Affect Measurement of Dynamic Properties of Clays

Dynamic properties of soils are dependent on a number of factors, such as OCR, effective stress, void ratio, and saturation (Athanasopoulos and Richart, 1983a, 1983b; Wu et al., 1984). The measurement of these dynamic properties is often influenced by strain rate effects (Isenhower and Stokoe, 1981). The strain rate effect of the measured variation of shear modulus with shear strain amplitude is shown in Figure 7.33.

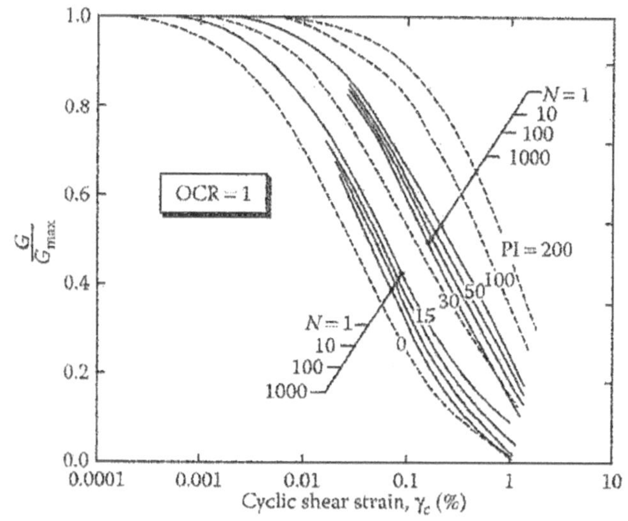

FIGURE 7.32 Effect of cyclic degradation on shear modulus (After Vucetric and Dobry, 1991). Reprinted with permission of ASCE.

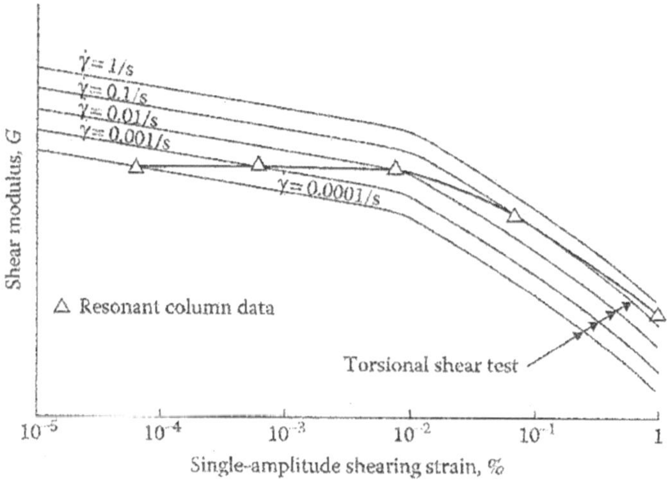

FIGURE 7.33 Combined strain amplitude and strain rate effects on shear modulus (After Isenhower and Stokoe, 1981).

Laboratory measurement of dynamic properties of soft soils are also often complicated by sample disturbance effects. Cuny and Fry (1973) reported +50% variation between laboratory and field measured values of soil moduli. Pore pressure build-up and strength deterioration accounts for part of the influences in laboratory

TABLE 7.3

Effect of Increase of Various Factors on *Grow* G/G_{max} and Damping Ratio A of Normally Consolidated and Moderately Over-consolidated Soils

Increasing Factor	G_{max}	G/G_{max}	D
Confining pressure σ	Increases with σ	Stays constant or increases with σ	Stays constant or decreases with σ
Void ratio, e	Decreases with e	Increases with e	Decreases with e
Geologic age, t	Increases with t	May increase with t	Decreases with t
Cementation, c	Increases with c	May increase with c	May decrease with c
Over-consolidation, OCR	Increases with OCR	Not affected	Not affected
Plasticity index, PI	Increases with PI if OCR > 1; stays about constant if OCR = 1	Increases with PI	Decreases with PI
Cyclic strain, γ_c	–	Decreases with γ_c	Increases with γ_c
Strain rate γ (frequency of cyclic loading)	Increases with strain rate	G increase with γ_c G/G_{max} probably not affected if G and G_{max} are measured at same γ	Stays constant or may increase with γ
Number of loading cycles, N	Decreases after N cycles of large γ_c but recovers later with time	Decreases after N cycles of large γ_c (G_{max} measured before N cycles)	Not significant for moderate γ_c and N

Source: Dobry and Vucetic (1987). Reprinted with permission of M. Vucetic.

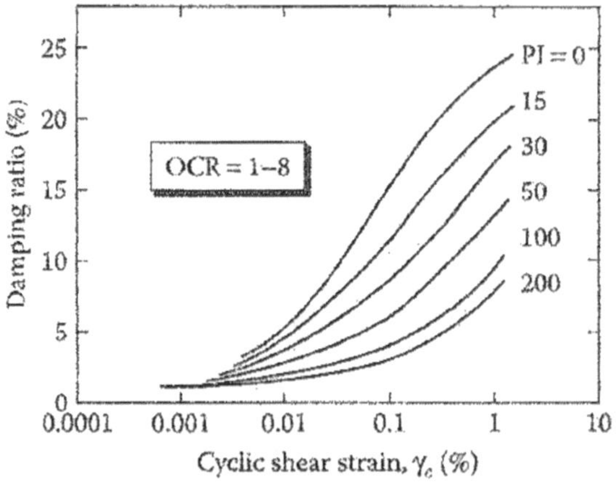

FIGURE 7.34 Variation of damping ratio of the fine-grained soil with cyclic shear strain amplitude and plasticity index (After Vucetric and Dobry, 1991). Reprinted with permission of ASCE.

measurements. Anderson and Woods (1975, 1976) have shown that shear wave velocity and therefore G increases approximately linearly with the logarithm of time after the end of primary consolidation. The change in shear modulus with time can be expressed as follows:

$$\Delta G_{max} = N_G \left(G_{max}\right)_{1000} \tag{7.19}$$

where

ΔG_{max} is the increase in G_{max} over one log cycle of time

$(G_{max})_{1000}$ is the value of G_{max} at a time of 1000 min after the end of primary consolidation

$$N_G = 0.027\sqrt{PI} \tag{7.20}$$

N_G has been shown to increase with increasing PI and decreases with increasing OCR (Kokushu et al., 1982). Lefebvre and LeBoeuf (1987) studied the effects of strain rate and cyclic loading on sensitive marine clays, in which they showed that reducing the strain rate or cycling the load appears to weaken the clay by a fatigue phenomenon due to reduction of effective stress and weakening of brittle bonds in the clay skeleton. They concluded that strain rate effects in saturated clays cause the cyclic strength mobilized at high frequencies to be higher than the monotonic strength measured at standard strain rates. Pamukcu (1989) has shown that for normally consolidated soft kaolinite clay, the ratio of the static to dynamic shear modulus, measured at a shearing strain amplitude of 10%–2%, is in the order of 0.85. Athanasopoulos and Richart (1983a) showed that at rates of deformation characterizing initial stages of creep, clays exhibited significant loss of stiffness. However, they regained this loss and exhibited increase over the pre-creep value at later stages of creep with slower rates of deformation. This was not the case for over-consolidated clays, in which the recovery and increase of stiffness did not occur with extended creep.

Macky and Saada (1984) developed empirical models relating shear modulus, shear strain, and consolidation pressure for cross-anisotropic clays. They reported that under slow cyclic loading the G/Gmax ratio decreased to about 50% of its original value at around 0.05% shear strain amplitude for the clays tested. Koutsoftas (1978) showed that the stiffness reduction was more significant than undrained strength reduction of two marine clays following cyclic loading. Athanosopoulos and Richart (1983b) showed that temporary release of confinement of cohesive soils caused a reduction in shear modulus. However, the initial value was regained when the confinement was reapplied over an interval of time. This modulus-regain time increased with the age of the cohesive soil. Temporary high-amplitude cyclic loading had a similar effect on soil moduli. These findings were suggested to explain some of the discrepancies between field and laboratory measurements of soil moduli. Ray and Woods (1988) showed that, with cyclic loading, the soil skeleton can change even without the generation of pore pressures. They conducted cyclic tests on dry sand and silt specimens and showed that silty soils exhibit reduction in shear modulus and damping ratios with the number of cycles of loading. Wu et al. (1984) studied the effect of saturation on sandy and silty soils, in which they found that capillary effects significantly increased the shear modulus of silty soils. For the specimens tested, the maximum increase in shear modulus (about two times the dry or fully saturated specimen) occurred between degrees of saturation of 5% and 20%. An empirical equation was developed to correct the shear modulus values measured at a particular degree of saturation for silty and sandy soils.

7.6 ANALYTICAL METHODS TO PREDICT CYCLIC RESPONSE OF CLAYS

There are a few analytical procedures that are used to estimate permanent strains resulting from cyclic loading of soils. Some of these approaches utilize hysteretic stress–strain relationships, and some use empirical relations (Bouckovalas et al., 1986). Prediction of permanent strains due to cyclic loading may be complicated in clays owing to the dependency of those strains on other factors, such as generation and dissipation of pore pressures and time involved in consolidation. Hyde and Brown (1976) suggested that permanent strain behavior of clay and silt soils under repeated loading and creep loading may have similarities. They demonstrated that, at first approximation, permanent strains accumulated during one-way cyclic loading can be estimated from static creep load test data. In evaluation of seismic response of soft clay sites it is more appropriate to use nonlinear analysis that equivalent linear methods. A nonlinear analysis incorporates the nonlinear stress–strain response and modulus degradation effects where an equivalent linear method utilizes strain-dependent stiffness and damping. Some of the nonlinear models of stress–strain behavior are bilinear, multilinear, hyperbolic and Ramberg Osgood idealizations. Singh et al. (1981) used a nonlinear model that incorporated modulus degradation developed by Idriss et al. (1978), on selected sites for seismic response analysis. A comparison of the acceleration and velocity spectra results from nonlinear analysis, equivalent linear analysis, and recorded values was made. Based on this study a good agreement was obtained between ground response and the nonlinear analysis response. The study also showed the significance of parametric studies involving possible variations in assumed rock motions and soil parameters when using response calculations. It was concluded that, at the levels of earthquake intensities where modulus degradation of soft clay sites becomes significant, the nonlinear analysis would yield more realistic results than equivalent linear methods.

Another example of evaluation of seismic response of cohesive soils was given by Tsai et al. (1980). It was concluded that nonlinear deformation, failure, and degradation behavior of the soil profile can have significant influence on seismic response under strong levels of earthquake shaking.

7.7 BEHAVIOR OF GRANULAR MATERIALS

If a relatively loose (relative density 70%), saturated sandy soil is subjected to cyclic loads it tends to compact and decrease in volume. If drainage is prohibited during the cyclic loading there is a subsequent buildup in pore water pressure until it is equal to the overburden pressure. The effective stress becomes zero and the soil loses its shear strength and develops a condition where it no longer can support shear loads (liquefied condition). The factors that influence cyclic loading-induced liquefaction are (1) soil type, (2) relative density or void ratio, (3) initial confining pressure, (4) nature of cyclic loading (magnitude and duration), and (5) level of saturation (Chaney, 1978).

A typical test record of a cyclic loading test on a sand material is shown in Figure 7.35. A review of Figure 7.35 shows that as the cyclic loading is occurring there is a corresponding buildup in the pore water pressure and a subsequent axial strain. If this condition is of a large extent, and the pore water pressure is not relieved, a lateral movement may result. The corrected stress ratio (i.e., cyclic stress ratio) as a function of the number of cycles of loading is shown for Monterey No. 0 sand at four relative densities in Figure 7.36. A review of Figure 7.36 shows that for increasing relative densities the number of load cycles to cause initial liquefaction increases.

Liquefaction is fundamentally controlled more by shear strain than by shear stress. This results from the generation of pore pressure due to the shear strain breaking down the soil structure and a corresponding tendency to densify. There is a level of shear strain, or threshold shear strain below which no pore pressure is generated. Using the volumetric strain (densification) that would occur if drainage were permitted and the slope of the rebound curve (E) the induced pore pressure can be determined as shown in the following equation.

$$\Delta u = E_r \Delta \varepsilon_{rd} \tag{7.21}$$

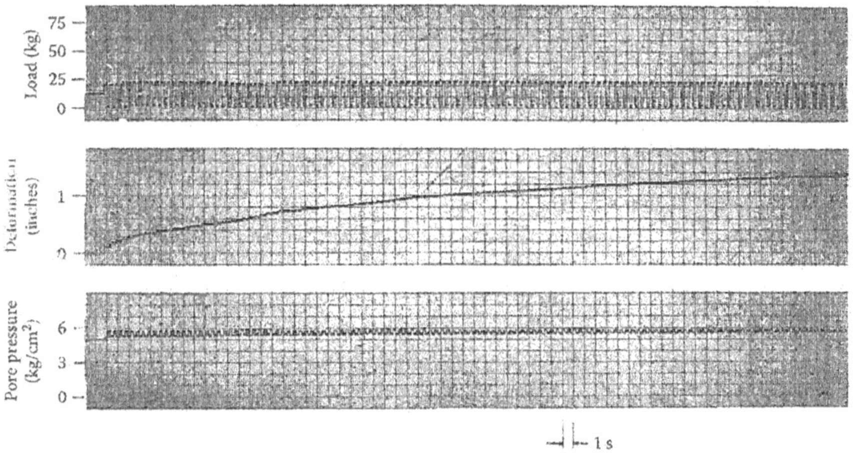

FIGURE 7.35 Record of typical non-reversing test on loose sand, K = 2.0, e = 0.87 (From Chaney and Fang, 1986). Reprinted with permission of ASTM.

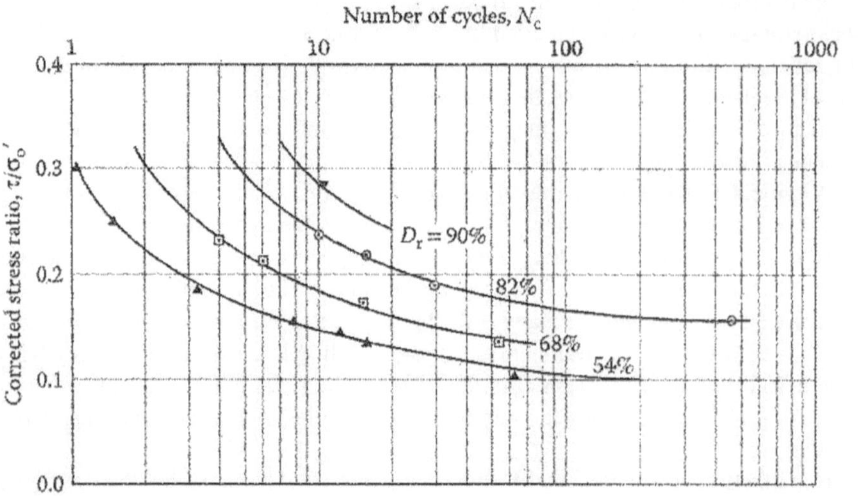

FIGURE 7.36 Corrected cyclic stress ratio and number of load cycles to cause initial liquefaction of sand at different initial relative densities (From De Alba et al., 1976). Reprinted with permission of ASCE.

Martin et al. (1975) have given procedures to evaluate these two parameters from the results of static rebound tests in a consolidation ring and cyclic load tests on dry sand, respectively. Finn and Bhatia (1981) have also reported good agreement between predicted and measured values using the proposed method. A relation has been presented by Martin et al. (1975) that relates the increase in residual pore water (Δu) for each load cycle to the rebound tangent modulus (E_r), porosity (n), bulk modulus of an air/water mixture (K_{aw}), and the volumetric strain per cycle ($\Delta\varepsilon_{vd}$), which is given by Equation 7.22.

$$\frac{\Delta u}{\Delta\varepsilon_{vd}} = \frac{1}{\left[\left(1/E_r\right)+\left(n/K_{aw}\right)\right]} \tag{7.22}$$

Assuming various values of and substituting a value of K_{aw} at atmospheric pressure is a qualitative measure of the effect of $\Delta\varepsilon_{vd}$ on pore pressure increment (Figure 7.37). A review shows that for saturation ($S = 100\%$) the residual pore pressure per cycle does not start to increase until a volumetric strain per cycle $\Delta\varepsilon_{vd}$ of $10^{-2}\%$.

7.7.1 Mechanism and Implications of Liquefaction Phenomena

The basic cause of liquefaction in saturated cohesionless soils is believed to be the buildup of excess hydrostatic pressure due to the application of either a shock or cyclic loading. An evaluation of the effect of saturation on liquefaction behavior as reflected by the Skempton B-value for a loose sand has been presented by Chaney (1978). It was shown in this study that as B value decreases (i.e., decreasing level of saturation), the number of cycles to liquefaction increases. As a consequence of the

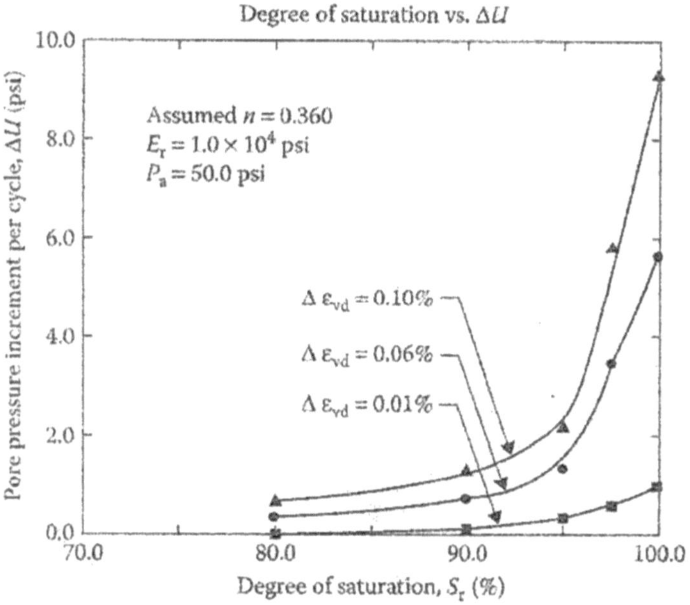

FIGURE 7.37 Change in pore pressure per load cycle as a function of degree of saturation (From Chaney, 1978). Reprinted with permission of ASCE.

applied stresses, the structure of the cohesionless soil tends to become more compact with a resulting transfer of stress to the pore water and a reduction in stress on the soil grains. The soil grain structure, in response, rebounds to the extent required to keep the volume constant and this interplay of volume reduction and soil structure rebound determines the magnitude of increase in pore water pressure in the soil (Martin et al., 1975). This behavior is slightly modified by the effect of stress concentrations (Chaney, 1980). As the pore water pressure approaches a value equal to the applied confining pressure, the sand begins to undergo deformations. If the sand is loose, the pore pressure will increase suddenly to a value equal to the applied confining pressure, and the sand will rapidly begin to undergo large deformations. If the sand will undergo virtually unlimited deformations without mobilizing significant resistance to deformation, it can be said to be liquefied. In contrast, if the sand is dense, it may develop a residual pore water pressure after completion of one full cycle of loading which is equal to the confining pressure. On application of the next cycle of loading, or if the sand is subject to monotonic loading, the soil will tend to dilate. The corresponding pore water pressure will drop if the sand is undrained, and the soil will develop enough resistance to with stand the applied load. In order for this to occur, the soil will have to undergo some deformation to develop the resistance. If the cyclic loading continues, the amount of deformation required to produce a stable condition may increase. Ultimately, there will be a level of deformation at which the soil can withstand any load application. This type of behavior is termed cyclic mobility and is considerably less serious than liquefaction. In both cases there is a generation of excess pore water pressure, which must be dissipated. The dissipation of the pore pressure can affect the behavior of overlying sediments by the formation of sand boils. The movement of pore water due to excess pressure results in the erosion and movement of sediments from the liquefied zones to the soil surface. This migration of sediments leaves voids in the underlying soil stratum, which ultimately collapses due to the overburden weight and thus, causes the distortion of the surface.

Due to the segregating action of wave attack on coastal materials and the resulting seaward transport, most coastal deposits have a relatively narrow particle size range as discussed by Fang and Chaney (1986). These relatively uniform deposits make them susceptible to quake liquefaction. In addition, the deposits void ratios normally exceed their critical void ratio (CVR) and, therefore, are in a potentially liquid state. These types of soils may be changed into actual macromeritic liquids (Winterkorn and Fang, 1975), throughout the whole granular system. A more detailed discussion of the liquefaction phenomena in the marine environment has been presented by Chaney and Fang (1986).

7.7.2 Variations in Liquefaction

There are two general classes of liquefaction phenomena depending on loading: (1) spontaneous liquefaction and (2) cyclic liquefaction. Spontaneous liquefaction is the result of the collapse of a metastable grain skeleton when subject to a mild shock. The mild shock can be generated by sudden differential settlement or because of a sudden rotational failure of a slope, or blasting effects.

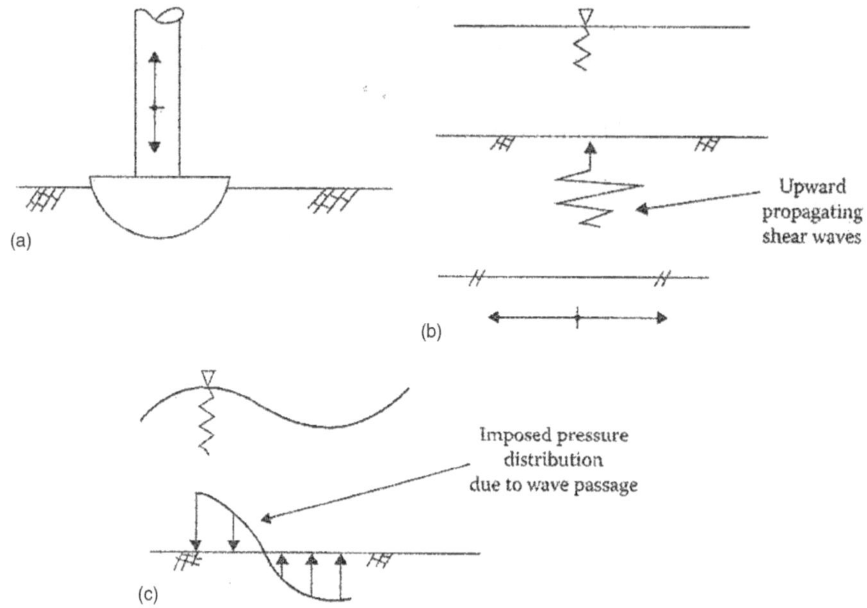

FIGURE 7.38 Typical types of cyclic loadings in the marine environment. (a) Jackup platform leg (1-way); (b) earthquake loading (2-way); (c) ocean wave loading (2-way) (From Chaney and Fang, 1991). Reprinted with permission of Taylor & Francis Group.

In contrast, cyclic liquefaction is the result of the application of either one-way (compressional) or two-way (compressional/tension) loadings (Figure 7.35). Cyclic stresses can be the result of earthquake loading conditions, wave loading of seabed sediments, movement of jack-up platform legs to wave loading, or sustained vibrations from heavy equipment and rail traffic. A summary of these type of loadings is shown in Figure 7.38.

7.8 WAVE INTERACTION WITH SEABED

7.8.1 Linear (Airy) Theory

The seafloor will be subjected to time varying pressure in the nearshore region where the depth of water is not great and the waves can be considered stable as shown in Figure 7.39. The relationship between wave period (*T*), wavelength (*L*), and water depth (*d*) for a rigid bottom is presented in Figure 7.40 based on the equations in Figure 7.39. Pressure on the rigid seabed due to passage of water waves can be determined using Airy linear wave theory and assuming the amplitude of waves is small relative to water depth and the seabed is rigid and impermeable. Based on Figure 7.39 the pressure is given by the following equation (7.23).

$$\Delta p = \Delta p' \sin 2\pi \left(\frac{x}{L} - \frac{t}{T} \right) \tag{7.23}$$

where

$$\Delta p' = \gamma_w \frac{H}{2} \left[\frac{1}{\cosh(2\pi d/L)} \right] \tag{7.24}$$

The value of *L* and *d* is obtained from the Airy wave theory.

$$L = \left(\frac{gT^2}{2\pi} \right) \tanh\left(\frac{2\pi d}{L} \right) \tag{7.25}$$

The corresponding wave profile is given by the following equation.

$$y_s = -\frac{H}{2} \sin 2\pi \left(\frac{x}{L} - \frac{t}{T} \right) \tag{7.26}$$

where

- H is the double amplitude of wave
- γ_ω is the unit weight of water L
- d is the water depth
- T is the wave period
- L is the wave length

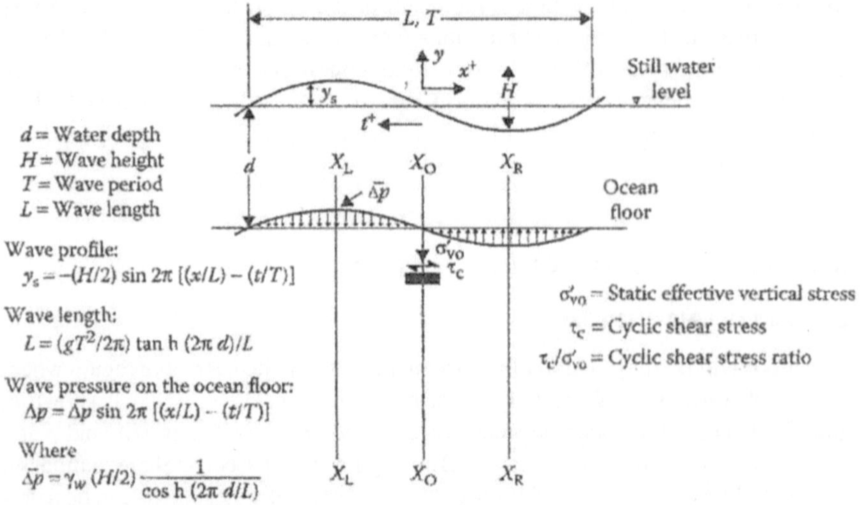

FIGURE 7.39 Relationships between wave profile, wavelength, and wave pressure on rigid seafloor. (From Seed and Rahman, 1978). Reprinted with permission of Taylor & Francis Group.

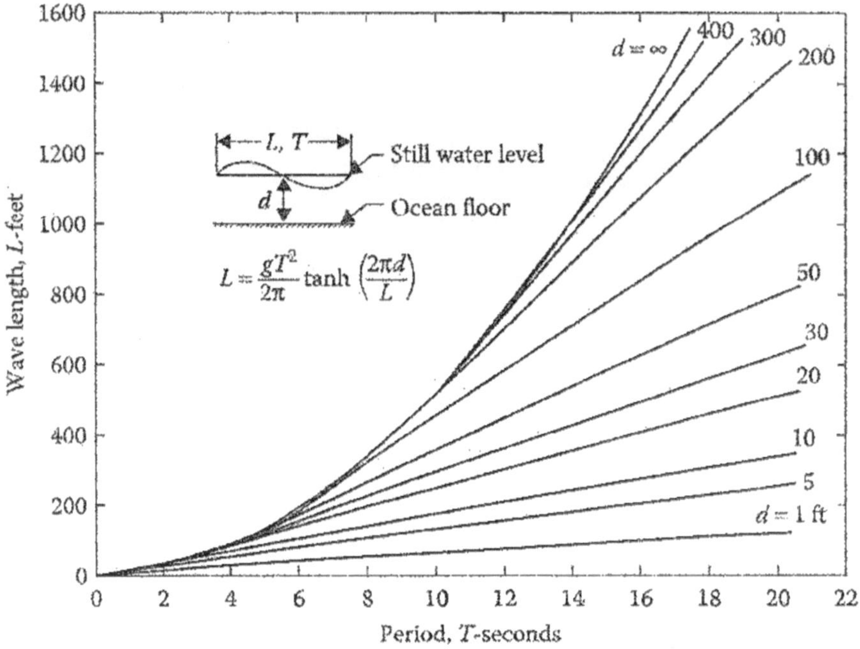

FIGURE 7.40 Relationship between wave period, wavelength, and water depth (From Seed and Rahman, 1978). Reprinted with permission of Taylor & Francis Group.

A review of this figure shows that there is a unique relationship between water depth, period, and wavelength. In particular, Figure 7.40 shows that wave-induced bottom pressures become important in a water depth less than approximately 152 m.

Waves in water depths less than 152 m cause a progressive increase in pore water pressure. The rate and amount of pore pressure buildup will depend on several factors: (1) height, period, and lengths of different wave components; (2) cyclic loading characteristics of the seabed deposits; and (3) drainage and compressibility of the soil deposit. The pressure Δp is given in Equation 7.19 and is plotted in Figure 7.41 as a dimensionless pressure amplitude against d/L. A review of Figure 7.41 shows that wave-induced bottom pressure are small when $d/L > 0.5$.

The general ocean wave problem differs from the earthquake problem in four major aspects:

1. The storm waves have periods considerably longer than earthquake loadings.
2. The duration of ocean storms is significantly longer than that of earthquake cyclic loadings.
3. There is a high probability that a structure in the ocean will be subjected to a number of minor storms followed by relatively quiet periods before the occurrence of the maximum design loading conditions corresponding to a 100-year storm.
4. Wave loading is at the mudline, whereas shear waves induced by earthquakes propagate upward from a lower level in the ground. In contrast, the earthquake problem assumes implicitly that
   a. The maximum earthquake will be the first and perhaps the only significant seismic disturbance to affect the site during the lifetime of the structure.
   b. Because of the very short duration of an earthquake, there could be no drainage of any excess pore water pressure developed during the cyclic loading.

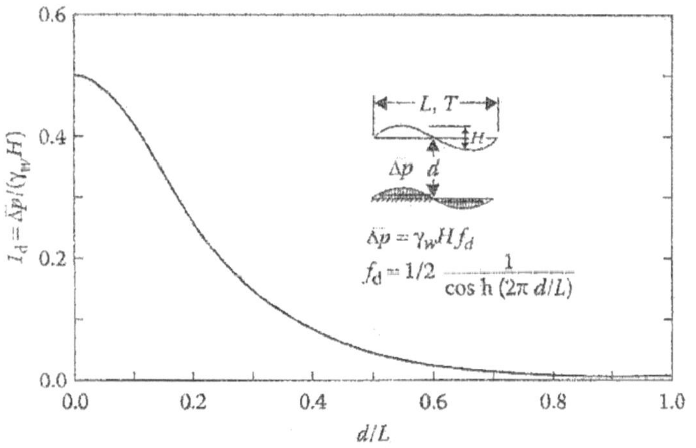

FIGURE 7.41 Wave-induced pressure on ocean floor (From Seed and Rahman, 1978). Reprinted with permission of Taylor & Francis Group.

The behavior of the soil following liquefaction depends on the type of cyclic loading. For earthquakes that produce upward propagating waves, there is some argument that after liquefaction, shear stress waves cannot be transmitted and, therefore, laboratory deformations following liquefaction in controlled stress cyclic loading tests may not be directly meaningful. However, the intensity of ocean wave-induced stresses would continue undiminished after liquefaction so that deformations would tend to be more significant for ocean waves than the earthquake problem.

The propagation of water waves over a permeable seabed exerts a time-varying pressure at the sediment/water interface. The time-varying pressure will cause cyclic variations in pore pressure and stresses within the bed. The effective stress varies in response to wave loading. Since soil strength is directly related to effective stress, any change in the effective stress state within the bed will affect bed strength and stability. Many coastal structures such as pipelines, platforms, anchors, and breakwaters that interact with the seabed will be affected by both cyclic effective stresses and the erosion potential of the bottom sediments. There have been a number of analytical studies that have examined the hydraulics of waves interacting with the seabed under a variety of conditions. Some of these conditions are the following:

- Saturated beds with isotropic permeability
- Saturated anisotropic permeability
- Stratified permeability
- Unsaturated or compressible seabed
- Effective stress state within bed.

The pressures calculated using the above approaches generally use the linear (Airy) theory and expect the pore pressure response to be in phase with the surface waves. This is in agreement with what has been measured for sandy bottoms. In contrast, measurements made by Hirst and Richards (1977) showed that bottom pressures can be much larger than predicted by linear theory for soft clayey bottoms. In Sections 7.8.2 and 7.8.3 the case of a rigid seabed (i.e., sandy bottom) will be considered first followed by the case of a deformable seabed (i.e., soft clayey bottom).

7.8.2 Rigid Seabed

The wave-induced pressure on a rigid ocean floor for a sinusoidal wave has been previously presented in Figure 7.41 (Seed and Rahman, 1978). The wave-induced pressures were shown to decrease with depth as given by Equation 7.23. These wind–wave-induced bottom pressures become important in water depths less than 152 m. A review of Figure 7.41 shows that the water pressure drops off exponentially.

Wave-induced shear stresses can be evaluated using the theory of elasticity. Simple charts for evaluating the shear stress ratio developed at any depth for waves having different characteristics are presented in Figure 7.42. The development of high pore pressures caused by the action of waves on an environment involving sand deposits

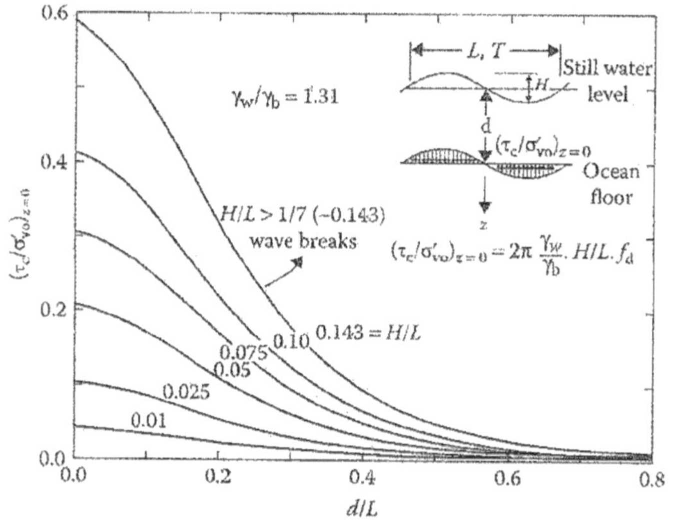

FIGURE 7.42 Relationship between wave characteristics and induced shear stress (From Seed and Rahman, 1978). Reprinted with permission of Taylor & Francis Group.

can lead to instability. This instability is an important concern for many engineering installations, such as pipelines and anchors. Seed and Rahman (1978) developed a procedure for evaluating the magnitude and distribution of wave-induced pore pressures in ocean floor deposits. The generation of excess pore pressure in terms of the number of cycles NL required to cause initial liquefaction under the given stress conditions has been presented by Seed and Booker (1977) in Equation 7.27.

$$\frac{u_g}{\sigma'_{vo}} = \left(\frac{2}{\pi}\right)\arcsin\left(x^{1/20}\right) \tag{7.27}$$

where

σ'_{vo} is the initial vertical effective stress
θ is the empirical constant, average value 0.7
$x = N/NL$

7.8.3 Deformable Seabed

Part of the SEASWAB project was to acquire direct measurements of wave action and seafloor response of marine clays under wave action. Suhayda (1977) reported on field measurements of bottom oscillations and wave characteristics in a study of the interaction of fine-grained sediments and surface waves (Figure 7.43). This was accomplished by using a wave staff, pressure transducers, and accelerometer in East Bay, Louisiana. This area has a fine-grained clay bottom. The bottom sediments

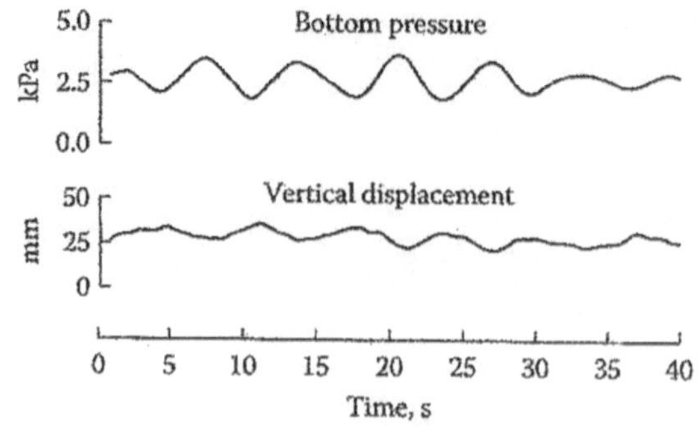

FIGURE 7.43 Simultaneous records of the wave-induced bottom pressure and the vertical displacement (From Suhayda, 1977). Reprinted with permission of Taylor & Francis Group.

appeared to be undergoing an elastic response to a surface wave crest. This was evidenced by the bottom being depressed by a surface wave crest. Under the range of bottom pressures measured, Suhayda (1977) reported that bottom displacement varied linearly with bottom pressures. The measured bottom pressures were found to be up to 35% larger than that predicted by using linear (Airy) wave theory. Therefore the energy lost from the surface wave to the bottom is significant and larger than the energy lost to bottom friction. This relationship is shown as the ratio of bottom pressure for a moving bottom to the bottom pressure without any motion (Airy wave) versus relative depth as a function of the $B = b/a$ in Figure 7.44. The value a is the maximum amplitude of the surface wave while b is the maximum amplitude of the

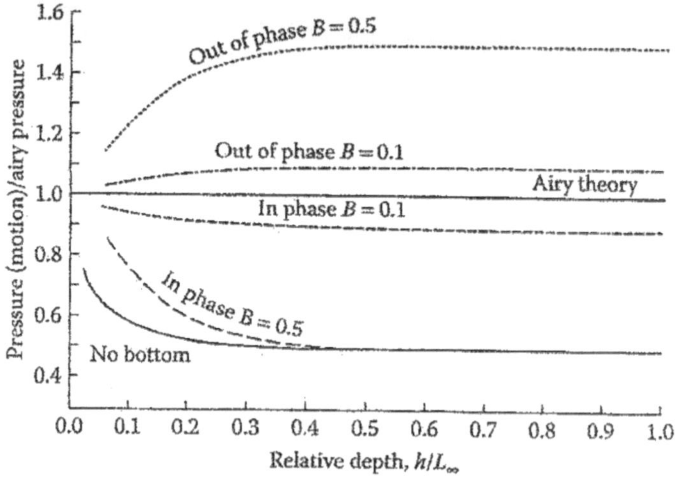

FIGURE 7.44 Ratio of the predicted bottom pressure over a moving bottom to the bottom pressure predicted from linear wave theory, as a function of relative depth. B is the ratio of b/a (From Suhayda, 1977). Reprinted with permission of Taylor & Francis Group.

bottom sediment motion. A review of this figure shows that for an out-of-phase wave an elastic-like response is generated resulting in a greater bottom pressure than if the waves are in phase. In summary, for reasonable wave heights, a flexible bottom dissipates energy at a rate at least an order of magnitude greater than a rigid impermeable bottom (Suhayda, 1977). A relatively greater amount of wave energy is lost on a muddy coast at intermediate water depths than is dissipated on a sandy coast (Suhayda, 1977).

7.9 POST-CYCLIC LOADING BEHAVIOR

The strength reduction due to cyclic loading has been studied by Theirs and Seed (1969), Castro and Christian (1976), Gardner (1977), Lee and Focht, 1976), Singh and Gardner (1979), Praeger and Lee (1978), Sangrey and France (1980), Van Eekelen and Potts (1978), Singh et al. (1978), and Anderson et al. (1980). The static strength behavior after cyclic loading based on these studies has been shown to depend on the drainage condition of the material prior to static loading, original OCR ratio, amount of strain generation, amount of pore water pressure generation., number of cycles of loading, the applied stress level, and the type of material. Results from undrained studies indicate that the cyclic loading produces an increase in pore pressure and a degradation of soil stiffness, which is reflected as a decrease in the effective stress and an associated increase in OCR_{max} as defined in the following equation:

$$OCR_{max} = \frac{\sigma'_{cm}}{\sigma'_c - \Delta u} \tag{7.28}$$

where

σ'_c is the existing consolidation stress

σ'_{cm} is the maximum past consolidation stress

Δu is the change in pore water pressure due to cyclic loading

The increase in pore pressure results in a decrease in shear strength. Van Eekelen and Potts (1978) derived a theoretical expression with critical state parameters, *k* and *l*:

$$\frac{S_{uc}}{S_u} - \left(1 - \frac{u_e}{\sigma'_c}\right)^{\kappa/\lambda} \tag{7.29}$$

where

S_{uc} is the postcyclic undrained shear strength

S_u is the precyclic undrained shear strength

u_e is the excess pore pressure due to cyclic loading

σ'_c is the initial effective confining pressure

k is the rebound or recompression index expressed on the natural logarithmic scale (=Cr/2.3)

l is the compression index expressed on the natural logarithm scale (=Cc/2.3)

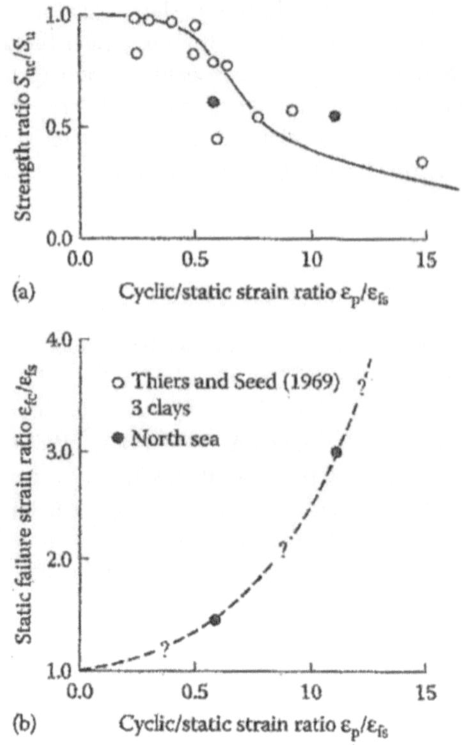

FIGURE 7.45 Effect of peak cyclic strain on monotonic strain after cyclic loading (after Lee and Focht, 1976). Reprinted with permission of Taylor & Francis Group.

Thiers and Seed have shown that the deterioration of undrained shear strength would be minimized when the cyclic strain is kept below one-half the pre-cyclic undrained shear strain at failure. The strength after cyclic loading is shown in Figure 7.45 (Lee and Focht, 1976).

REFERENCES

AnhDan, L., Tatsuoka, F., and Koseki, J. (2006). "Viscous effects on the stress-strain behavior of gravelly soil in drained triaxial compression," *Geotechnical Testing Journal*, 29(4): 330–340.

Anderson, D.G., Phukunhaphan, A., Douglas, B.J., and Martin, G.R. (1983). Cyclic behavior of six marine clays. Evaluation of seafloor soil properties under cyclic loads, *Proceedings of Session No. 52 of ASCE Annual Convention*, Houston, TX, October, 27pp.

Anderson, K.H., Pool, J.H., Brown, S.F., and Rosenbrand, W.F. (1980). "Cyclic and statis laboratory tests on drammen clay," *Journal of the Geotechnical Engineering Division, ASCE*, 106(GT-5): 499–529.

Anderson, D.G., and Woods, R.D. (1975). "Comparison of field and laboratory shear moduli," *Proceedings of the Conference on In-Situ Measurement of Soil Properties*, Specialty Conference of the Geotechnical Division, North Carolina State University, Raleigh, Vol. *I*, pp. 69–92.

Anderson, D.G., and Woods, R.D. (1976). "Time-dependent increase in shear modulus of clay," *Journal of the Geotechnical Engineering Division, ASCE*, 102(5): 525–537.

Ansal, A.M., and Erken, A. (1989). "Undrained behavior of clay under cyclic shear stresses," *Journal of the Geotechnical Engineering Division, ASCE*, 115(7): 968–983.

Athanasopoulos, G.A., and Richart, F.E. (1983a). "Effect of creep on shear modulus of clays," *Journal of Geotechnical Engineering, ASCE*, 109(10): 1217–1232.

Athanasopoulos, G.A., and Richart, F.E. (1983b). "Effect of stress release on shear modulus of clays," *Journal of Geotechnical Engineering, ASCE*, 109(10): 1233–1245.

Augustesen, A., Liingaard, M., and Lade, P.V. (2004). "Evaluation of time dependent behavior of soils," *International Journal of Geomechanics*, 4(3): 137–156.

Azzouz, A.S., Malek, A.M., and Baligh, M.M. (1989). "Cyclic behaviour of clays under undrained simple shear," *Journal of the Geotechnical Engineering Division, ASCE*, 112(6): 579–593.

Bouckovalas, G., Marr, W.A., and Christian, J.T. (1986). "Analyzing permanent drift due to cyclic loads," *Journal of the Geotechnical Engineering Division, ASCE*, 112(6): 579–593.

Byrne, P.M. (1988). Pers. comm. to Prof. H.G. Poulos, as presented in Poulos, H.G., *Marine Geotechnics*, Chap. 3, Unwin Hyman, London.

Castro, G., and Christian, J.T. (1976). "Shear strength of soils and cyclic loading," *Journal of the Geotechnical Engineering Division, ASCE* 102(9): 887–894.

Chaney, R.C. (1978). Saturation effects on the cyclic strengths of sand. In *Proceedings, Speciality Conference on Earthquake Engineering and Soil Dynamics*, ASCE, Pasadena, CA, Vol. 1, pp. 342–358.

Chaney, R.C. (1980). Seismically induced deformations in earthdams. In *7th World Conference on Earthquake Engineering*, Istanbul, Turkey, Vol. 31, pp. 483–486.

Chaney, R.C. (1984). Methods of predicting the deformation of the seabed due to cyclic loading. In *Proceedings of the Symposium on Seabed Mechanics, International Union of Theoretical and Applied Mechanics*, University of Newcastle, Callaghan, NSW, Australia, pp. 159–167.

Chaney, R.C. (2013). "Dynamic properties of some Eastern Mediterranean marine sediments," *Geotechnical Testing Journal, ASTM*, 36(4): 524–532.

Chaney, R.C., and Fang, H.Y. (1984). Cyclic degradation models of soils. In *Proceedings of the 5th ASCE—Engineering Mechanics Division Specialty Conference*. ASCE Press, New York, NY, 5pp.

Chaney, R.C., and Fang, H.Y. (1991). "Liquefaction in the coastal environment: An analysis of case histories," *Marine Geotechnology*, 10(3–4): 343–370.

Chaney, R.C., and Fang, H.Y. (1986). Static and dynamic properties of marine sediments. In *Proceedings of Symposium on Marine Geotechnology and Nearshore/Offshore Structures*, Shanghai, China, ASTM STP 923, pp. 74–111.

Christensen, R.W., and Wu, T.H. (1964). "Analysis of clay deformation as a rate process," *Journal of Soil Mechanics and Foundation Division*, 90(6): 125–157.

Cuny, R.W., and Fry, Z.B. (1973). "Vibratory in-situ and laboratory soil moduli compared," *Journal of the Geotechnical Engineering Division, ASCE*, 99(12): 1055–1076.

De Alba, P., Seed, H.B., and Chen, C.K. (1976). "Sand liquefaction in large-scale simple shear tests," *Journal of the Geotechnical Engineering Division, ASCE*, 102(GT9): 909–907.

Di Benedetto, H., Tatsuoka, F., and Ishihara, M. (2002). Time-dependent shear deformation characteristics of sand and their constitutive modeling. *Soils Foundations*, 42(2): 1–22.

Dobry, R., and Ladd, R.S. (1980). "Discussion on soil liquefaction and cyclic mobility evaluation for level ground during earthquakes. By H.B. Seed and Peck, R.B. Liquefaction potential: science versus practice," *Journal of the Geotechnical Engineering Division*, 106(6), 720–724.

Dobry, R., and Vucetic, M. (1987). Dynamic properties and seismic response of soft clay deposits. In *Proceedings of the International Symposium on Geotechnical Engineering of Soft Soils* 2, ISSMFE, Mexico City, Mexico, pp. 51–87.

Dobry, R. et al. (1982). *Prediction of Porewater Pressure Buildup and Liquefaction of Sands during Earthquakes by the Cyclic Strain Method*, NBS Building Science Series 138, US Department of Commerce, National Bureau of Standards, Washington, DC.

Drnevich, V.P., and Richart, F.E. Jr. (1970). "Dynamic prestraining of dry sand," *Journal of the Soil Mechanics and Foundation Division, ASCE* 96(2): 453–469.

Egan, J.A., and Sangrey, D.A. (1978). A critical state model for cyclic load pore pressure. In *Proceedings, ASCE Special Conference on Earthquake Engineering and Soil Dynamics*, Pasadena, CA, Vol. 1, pp. 411–424.

Fang, H.Y., Chaney, R.C., and Pandit, N.S. (1981). Dynamic shear modulus of soft silt. In *International Conference on Recent Advances in Geotechnical Earthquake Engineering and Soil Dynamics*, University of Missouri, Rolla, MO, pp. 575–580.

Fang, H.Y., and Chaney, R.C. (1986). Geo-environmental and climatological conditions related to coastal structural design along the China coastline. In *Proceedings of the Symposium on Marine Geotechnology and Nearshore/Offshore Structures*, Shanghai, China, ASTM, STP 923, pp. 149–160.

Feda, J. (1989). "Interpretation of creep of soils by rate process theory," *Geotechnique*, 39(4): 667–677.

Finn, W.D.L., and Bhatia, S.K. (1981). Prediction of seismic pore-water pressures. In *Proceedings, of the 10th International Conference on Soil Mechanics and Foundation Engineering*, Rotterdam, the Netherlands, Vol. 3, pp. 201–206.

France, J.W., and Sangrey, D.A. (1977). "Effects of drainage in repeated loading of clays," *Journal of the Geotechnical Engineering Division*, 103(6): 769–785.

Gardner, W.S. (1977). Soil property characterization in geotechnical engineering practice. Third Woodward Lecture. Geotechnical Environmental Bulletin 10/2. Woodward-Clyde Consultants, San Francisco, CA.

Glasstone, S., Laidler, K.J., and Eyring, H. (1941). "The theory of rate processes," *Journal of Chemical Education*, 19(5): 249.

Goulois, A.M., Whitman, R.V., and Hoeg, K. (1985). Effects of sustained shear stresses on the cyclic degradation of clay, *Proceedings of the Symposium on Strength Testing of Marine Sediments*, ASTM STP 883, pp. 336–351.

Hardin, B.O. (1978). "The nature of stress-strain behavior for soils," *Earthquake Engineering and Soil Dynamics*. ASCE, 1: 3–90.

Hardin, B.O., and Drnevich, V.P. (1972). "Shear modulus and damping in soils: Design equations and curves," *Journal of the Soil Mechanics and Foundations Division, ASCE*, 98(7): 667–692.

Herrmann, H.G., and Houston, W.N. (1976). Response of seafloor soils to combined static and cyclic loading. In *Proceedings of the Offshore Technology Conference*. Houston, TX, pp. 53–59.

Herrmann, H.G., and Houston, W.N. (1978). Behavior of the seafloor soils subjected to cyclic loading. In *Proceedings of the Offshore Technology Conference*, Houston, TX, Vol. 3, pp. 1797–1804.

Hirst, T.J. and Richards, A.F. (1977). "In situ pore-pressure measurement in Mississippi delta front sediments," *Marine Geotechnology*, 2: 191–204.

Hirst, T.J. (1968a). The influence of compositional factors on the stress-strain-time behavior of soils. PhD thesis, University of California, Berkeley, CA.

Hirst, T.J. (1968b). Compositional and environment influences on stress-strain-time behavior of soils. PhD thesis, University of California, Berkeley, CA.

Hyde, A.F.L., and Brown, S.F. (1976). "The plastic deformation of silty clay under creep and repeated loading," *Geotechnique*, 26(1): 173–184.

Idriss, I.M., Dobry, R., and Singh, R.D. (1978). "Nonlinear behavior of soft clays during cyclic loading," *Journal of the Geotechnical Engineering Division, ASCE*, 104(12): 1427–1447.

Isenhower, W.M., and Stokoe II, K.H. (1981). Strain-rate dependent shear modulus of San Francisco Bay mud. In *International Conference on Recent Advances in Geotechnical Earthquake Engineering and Soil Dynamics*, University of Missouri, Rolla, MO, pp. 597–602.

Ishihara, K., and Yasuda, S. (1980). "Cyclic strengths of undisturbed cohesive soils of western Tokyo," in A.A. Swansea (Ed.), *International Symposium on Soils Under Cyclic and Transient Loading*, Balkema, Rotterdam, the Netherlands, pp. 57–66.

Kokusho, T., Yoshida, Y., and Esashi, Y. (1982). Dynamic properties of soft clay for wide strain range. *Soils and Foundations*, 22: 1–18. 10.3208/sandf1972.22.4_1.

Koutsoftas, D.C. (1978). "Effect of cyclic loads on undrained strength of two marine clays," *Journal of Geotechnical Engineering, ASCE*, 104(5): 609–620.

Kramer, S.L. (1996). *Geotechnical Earthquake Engineering*. Prentice Hall, Upper Saddle River, NJ, 653pp.

Kuhn, M.R., and Mitchell, J.K. (1992). "New perspectives on soil creep". *Journal of Geotechnical Engineering*, 119(3): 507–524.

Kuwano, R., and Jardine, R. (2002). "On measuring creep behavior in granular materials through tri-axial testing," *Canadian Geotechnical Journal*, 39(5): 1061–1074.

Lade, P.V., Liggio Jr., C.D., and Jungman, N. (2009). "Strain rate, creep, and stress-drop-creep experiments on crushed coral sand," *Journal of Geotechnical and Geoenvironmental Engineering*, 135(7): 941–963.

Lee, K.L., and Focht, J.A. (1976). "Strength of clay subjected to cyclic loading," *Marine Geotechnology*, 1: 165–185.

Lefebvre, G., and LeBoeuf, D. (1987). "Rate effects and cyclic loading of sensitive clays," *Journal of the Geotechnical Engineering, ASCE*, 113(5): 476–489.

Liingaard, M., Augustesen, A., and Lade, P.V. (2004). "Characterization of models for time-dependent behavior of soils," *International Journal of Geomechanics*, 4(3): 157–177.

Macky, T.A., and Saada, A.S. (1984). "Dynamics of anisotropic clays under large strains," *Journal of Geotechnical Engineering, ASCE*, 110(4): 487–504.

Martin, G.R., Finn, W.D.L., and Seed, H.B. (1975). "Fundamentals of liquefaction under cyclic loading," *Journal of the Geotechnical Engineering Division, ASCE*, 101(GT-5): 423–438.

Masing, G. (1926). Eigenspannungen and Verfestigung beirn Messing. In *Proceedings of the 2nd International Congress of Applied Mechanics*, Corfu, Greece.

Matsui, T., Ohara, H., and Ito, T. (1980). "Cyclic stress-strain history and shear characteristics of clay," *Journal of the Geotechnical Engineering Division, ASCE*, 106(10): 1011–1020.

Matsushita, M., Tatsuoka, F., Koseki, J., Cazacliu, B., Benedetto, H., and Yasin, S.J.M. (1999). "Time effects on the pre-peak deformation properties of sands," in M. Jamiolkowski, R. Lancelotta, and D. Lo Presti (Eds.), *Pre-Failure Deformation Characteristics of Geo-Materials*, Balkema, Rotterdam, the Netherlands, pp. 681–689.

Meimon, Y., and Hicher, P.Y. (1980). "Mechanical behavior of clays under cyclic loading," In A.A. Swansea (Ed.), *Proceedings of the International Symposium on Soils Under Cyclic and Transient Loading*, Balkema, Rotterdam, the Netherlands, pp. 77–88.

Mindlin, R.D., and Deresiewicz, H. (1953). "Elastic spheres in contact under varying oblique forces," *Journal of Applied Mechanics, ASME*, 20: 327–344.

Mitchell, J.K., Campanella, R.G., and Singh, A. (1968). "Soil creep as a rate process," *Journal of the Soil Mechanics and Foundation Division*, 94(1): 231–253.

Mitchell, J.K. (1964). "Shearing resistance of soils as a rate process," *Journal of the Soil Mechanics and Foundation Division*, 90(1): 29–61.

Mitchell, J.K. (1976). *Fundamentals of Soil Behavior*. Wiley, New York, pp. 247–250.

Mitchell, J.K. (1993). *Fundamentals of Soil Behavior*, 2nd edn. Wiley, New York, 437pp.

Moses, G.G., and Rao, S.N. (2003). "Degradation in cemented marine clay subjected to cyclic compressive loading," *Marine Georesources and Geotechnology*, 21(1): 37–62, Taylor & Francis Group.

Murayama, S., and Shibata, T. (1964). Flow and stress relaxation of clays (Theoretical studies on the rheological properties of clay—part 1). In *Rheology and Soil Mechanics. Symposium of the International Union of Theoretical and Applied Mechanics*, Grenoble, France.

Pamukcu, S. (1989). "Shear modulus of soft marine clays," *Journal of Offshore Mechanics and Arctic Engineering, ASME*, 111(4): 265–272.

Pamukcu, S., Poplin, J.K., Suhayda, J.N., and Tumay, M.T. (1983). Dynamic sediment properties, Mississippi Delta. In *Proceedings of the Geotechnical Conference in Offshore Engineering*, ASCE, New York, pp. 111–132.

Pamukcu, S., and Suhayda, I. N. (1984). "Evaluation of shear modulus for soft marine clays, Mississippi Delta," in R.C. Chaney and K.R. Demars (Eds.), *Strength Testing of Marine Sediments: Laboratory and In-Situ Measurements*, STP 883, ASTM, New York, pp. 352–362.

Pamukcu, S., and Suhayda, J.N. (1987). "High resolution measurement of shear modulus of clay using tri-axial vane device," in A.S. Carmak (Ed.), *Soil Dynamics and Liquefaction*, Developments in Geotechnical Engineering No. 42, Elsevier and Computational Mechanics Publications, Amsterdam, pp. 307–321.

Park, T.K., and Silver, M.L. (1975). "Dynamic triaxial and simple shear behavior of sand," *ASCE Journal of the Geotechnical Engineering Division* 101(6): 513–529.

Poulos, H.G. (1988). *Marine Geotechnics*, Chap. 3. Unwin Hyman, London.

Poulos, S.J. (1981). "The steady state of deformation," *Journal of the Geotechnical Engineering Division*, 107(5): 553–562.

Praeger, S.R., and Lee, K.L. (1978). "Post-cyclic strength of marine limestone soils," in *Earthquake Engineering and Soil Dynamics*. ASCE Press, New York, Vol. 1, pp. 732–745.

Prevost, J.H. (1977). "Mathematical modelling of monotonic and cyclic undrained clay behavior," *International Journal of Numerical and Analytical Methods in Geotechnical Engineering*, 1(2): 195–216.

Procter, D.C., and Khaffaf, J.H. (1984). "Cyclic triaxial tests on remoulded clay," *Journal of Geotechnical Engineering, ASCE*, 110(10): 1431–1445.

Pyke, R. (1973). Settlement and liquefaction of sands under multi-directional loading. PhD dissertation, University of California, Berkeley, CA.

Ramberg, W., and Osgood, W.T. (1943). Description of stress-strain curves by three parameters. Technical Note 902. NASA, Washington, DC.

Ray, R.P., and Woods, R.D. (1988). "Modulus and damping due to uniform and variable cyclic loading". *Journal of Geotechnical Engineering, ASCE*, 114(8): 861–876.

Sangrey, D.A. (1977). "Marine geotechnology—State of the art," *Marine Geotechnology* 2: 45–80.

Sangrey, D.A., Castro, G., Poulos, S.J., and France, J.W. (1978). "Cyclic loading of sands, silts and clays," in *Proceedings, ASCE Special Conference on Earthquake Engineering and Soil Dynamics*, Pasadena, CA, Vol. 2, pp. 836–851.

Sangrey, D.A., and France, J.W. (1980). "Peak strength of clay soils after a repeated loading history," in *International Symposium on Soils Under Cyclic and Transient Loading*, Swansea, A.A., ed. Balkema, Rotterdam, the Netherlands, Vol. *I*, pp. 421–430.

Sangrey, D.A., Henkel, D.J., and Esrig, M.I. (1969). "The effective stress response of a saturated clay soil to repeated loading," *Canadian Geotechnical Journal*, 6(3): 241–252.

Schiffman, R.L. (1959). *The Use of Visco-Elastic Stress-Strain Laws in Soil Testing*. ASTM. Spec. Tech. Pub. No. 254. ASTM, Philadelphia, PA, pp. 131–155.

Seed, H.B., and Booker, J.R. (1977). "Stabilization of potentially liquefiable sand deposits using gravel drains," *Journal of the Geotechnical Engineering Division, ASCE*, 103/7: 757–768.

Seed, H.B., and Chan, C.K. (1966). "Clay strength under earthquake loading conditions," *Journal of the Mechanics and Foundations Division, ASCE*, 92(2): 53–78.

Seed, H.B., and Rahman, M.S. (1978). "Wave-induced pore pressure in relation to ocean floor stability of cohesionless soils," *Marine Geotechnology*, 3(2): 123–150.

Singh, A. (1966). Creep phenomena in soils. PhD thesis, University of California, Berkeley, CA.

Singh, A., and Mitchell, J.K. (1968). "A general stress-strain-time function for soil," *Journal of Soil Mechanics and Foundations Division*, 94(1): 21–46.

Singh, R.D., and Gardner, W.S. (1979). "Characterization of dynamic properties of Gulf of Alaska clays," in *Proceedings of the Soil Dynamics in Marine Environment*. Preprint 3604. ASCE, New York, 13pp.

Singh, R.D., Kim, J.H., and Caldwell, S.R. (1978). "Properties of clays under cyclic loading," in *Proceedings of the 6th Symposium on Earthquake Engineering*, University of Roorkee, Uttarakhand, India, 107–112.

Singh, R.D., Ricardo, D., Doyle, E.H., and Idriss, I. M. (1981). "Nonlinear seismic response of soft clay sites," *Journal of the Geotechnical Engineering Division, ASCE*, 107(GT-9): 1201–1218.

Stokoe II, K.H. (1980). Dynamic properties of offshore silty samples. In *12th Annual Offshore Technology Conference*, OTC 3 77 1, Houston, TX, pp. 289–295.

Suhayda, J.N. (1977). "Surface waves and bottom sediment response," *Marine Geotechnology*, 2: 135–146.

Tatsuoka, F., Santucci de Magistris, F., Hayano, K., Momoya, Y., and Koseki, J. (2000). "Some new aspects of time effects on the stress-strain behavior of stiff geomaterials," in R. Evangelista and L. Picarelli (Eds.), *The Geomechanics of Hard Soils-Soft Rocks*, Balkema, Rotterdam, the Netherlands, pp. 1285–1371.

Tatsuoka, F., Shihara, M., Di Benedetto, H., and Kuwano, R. (2002). "Time-dependent shear deformation characteristics of geomaterials and their simulation," *Soils and Foundations*, 42(2): 103–129.

Tatsuoka, F., Enomoto, T., and Kiyota, T. (2006). "Viscous property of geomaterial in drained shear. Geomechanics II—Testing, Modeling and Simulation," in P.V. Lade and T. Nakai (Eds.), *Proceedings of the 2nd Japan-US Workshop*, ASCE Geotechnical Special Publication No. 156, ASCE Press, New York, pp. 285–312.

Tavenas, F., Lerouell, S., La Rochelle, P., and Roy, M. (1978). "Creep behavior of an undisturbed lightly overconsolidated clay," *Canadian Geotechnical Journal*, 15(3): 402–423.

Thiers, R.G., and Seed, H.B. (1969). *Strength and Stress-strain Characteristics of Clays Subjected to Seismic Loading Conditions, Vibration Effects of Earthquakes on Soils and Foundations*, ASTM STP 450, Philadelphia, PA, pp. 3–56.

Togrol, E., and Guler, E. (1984). "Effect of repeated loading on the strength of clay," *Soil Dynamics and Earthquake Engineering*, 3(4): 184–190.

Tsai, C.F., Lam, L., and Martin, G.R. (1980). "Seismic response of cohesive marine soils," *Journal of the Geotechnical Engineering Division, ASCE*, 106(GT-9): 997–1012.

Vaid, Y., and Campanella, R.G. (1977). "Time-dependent behavior of undisturbed clay," *Journal of the Geotechnical Engineering Division*, 103(7): 693–709.

Van Eekelen, H.A.M., and Potts, D.M. (1978). "The behavior of drammen clay under cyclic loading," *Geotechnique* 8(2): 173–196.

Vucetic, M. (1988). "Normalized behavior of offshore clay under uniform cyclic loading," *Canadian Geotechnical Journal*, 25(1): 33–41.

Vucetric, M. (1994). "Cyclic threshold shear strains in soils," *ASCE Journal of Geotechnical Engineering*, 120(12): 2208–2228.

Vucetric, M., and Dobry, R. (1991). "Effect of soil plasticity on cyclic response," *Journal of Geotechnical Engineering, ASCE*, 117(1): 89–107.

Vyalov, S.S. (1986). *Rheological Fundamentals of Soil Mechanics*. Elsevier, New York, 564pp.

Winterkorn, H.F., and Fang, H.Y. (1975). "Soil technology and engineering properties of soils," in H.F. Winterkorn and H.Y. Fang (Eds.), *Foundation Engineering Handbook*, Van Nostrand-Reinhold, New York, pp. 91–116.

Wu, S., Gray, D., and Richart Jr, F.E. (1984). “Capillary effects on dynamic modulus of sands and silts,” *Journal of the Geotechnical Engineering Division, ASCE*, 110(9): 1188–1203.

Youd, T.L. (1972). “Compaction of sands by repeated shear straining,” *ASCE Journal of the Soil Mechanics and Foundations Division*, 98(7): 709–725.

Zhu, J.-G., Yin, J.-H., and Luk, S.-T. (1999). “Time-dependent stress-strain behavior of soft Hong Kong Marine Deposits,” *Geotechnical Testing Journal*, 22(2): 112–120.

8 Sedimentation

8.1 INTRODUCTION

Sedimentation is defined as the process of depositing sediments. This includes the separation of rock particles from their origin, the transportation of these particles, their site of deposition, and all changes and stages leading to the eventual consolidation of the sediment into sedimentary rock. The type of sediments found in marginal seas and deep ocean environments is dependent on their proximity to land, weather patterns, and depth.

The major sedimentologic features in the South China Sea (SCS) and the Taiwan Strait will be discussed in the following. Terrigenous sediments include fluvial sediments (modern or relict) that form the major river deltas, mud fields, and sand ridges on the shelf. In biogenic sediments, the distributions of coral reefs, calcareous and siliceous biogenic components, and organic carbon will be presented, followed by volcanic sediments in the SCS deep sea.

At present, the four largest rivers in the region: Yangtze, Pearl, Red, and Mekong transport terrestrial sediment into the basins comprising the SCS as shown previously in Figure 2.14.

The China Seas (South China Sea and the Taiwan Strait are the southern part of the China Seas) marginal basins capture the majority of land-derived sediments. This process produces the very high sedimentation rates inside the basins as compared to very low values in the open Pacific. The low sedimentation rate explains the lack of large deep-sea fans from the western Pacific. In contrast, large deep-sea fans do occur in areas of high sedimentation rates such as the Indian (Bengal and Indus) and Atlantic (Amazon). The high sedimentation rates coupled with marginal basins are related to the existence of wide continental shelves. This is especially true with the Sunda Shelf and the East China Sea shelf, as a large-scale coastline migration over glacial cycles had caused drastic changes in regional sedimentology.

8.2 SHELF SEDIMENTATION PATTERNS

Sediments in the SCS and Taiwan Straits are predominantly terrestrial in nature on the shelves (i.e., sand, silt, and clays). This is because of weathering from nearby land masses due to erosion. Clay and sand that are carried by the surrounding rivers constitute the main part of the sediments deposited on the seafloor. In addition, there are small amounts of calcareous materials. Sediment types common in the China Seas are sand, silt, and silty clay on the shelf and increasing biogenic components in the deep sea basin as shown in Figures 8.1 and 8.2.

Figure 8.1 concentrates on sediments close to the People's Republic of China coastline. A review of Figure 8.1 shows a narrow shelf bordered by a steep irregular

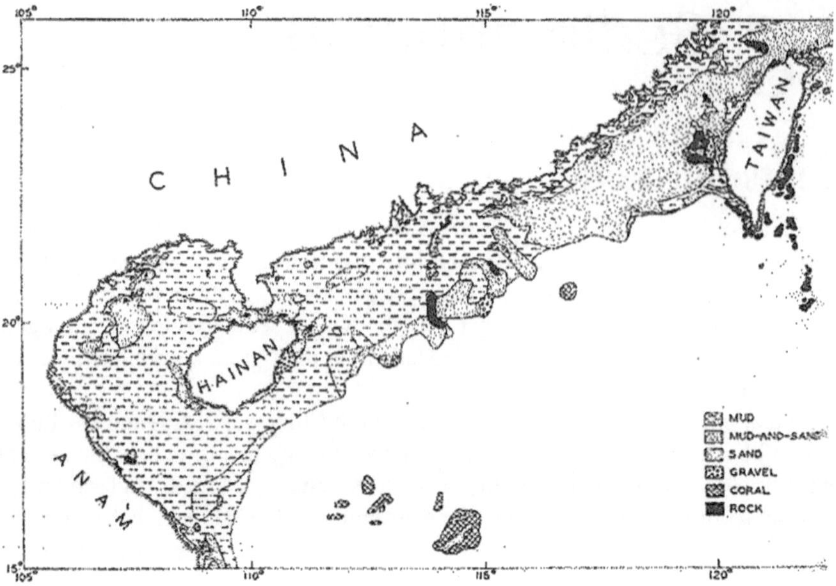

FIGURE 8.1 Sedimentation patterns along the Chinese coast and the Taiwan Strait (Shepard, 1973). Public domain.

slope along the east coast of Taiwan. In the Taiwan Strait, the sediment is largely relict sand. Continuing along the Chinese coast the shelf sediments as far south as Hong Kong are mainly considered predominately relict. South of Hong Kong, large rivers contribute recent muds, which form the sediment cover of the shelf except for narrow sand belts where the Kuroshio current has prevented deposition. Beyond this wide shelf that forms the Gulf of Tonkin, the shelf narrows and becomes very irregular along the coast of South Vietnam. South-west of Hainan, a large group of coral banks lie east of a deep basin, with depths of approximately 1500 m. The gentle continental slopes contrast greatly with the steep slopes that border the shelves farther south.

In contrast, Figure 8.2 presents the area further south. The Sunda shelf and Gulf of Thailand comprise one of the world's widest continental shelves. The broad shallow sea has been studied by a number of investigators including Emery and Niino (1963). They reported that the northern part of the Sunda shelf is a geosyncline that was formed during the late Tertiary and contains thick sediments. The shelf sediments are mostly mud, with a broad band of relict sand with reef patches on the outside. The north-flowing Kuroshio current sends an arm up the east side of the Gulf with a return south flow to the west. A broad basin with depths of over 80 m occurs in the inner Gulf.

Two other sediment-filled basins were found in the SCS by Parke et al. (1971). One basin extends into southern South Vietnam and along the Vietnamese coast to the north-east. This basin is separated by a sill from a larger basin extending east to north-west of Borneo. Sediments in all the basins are at least 2 km thick, are folded,

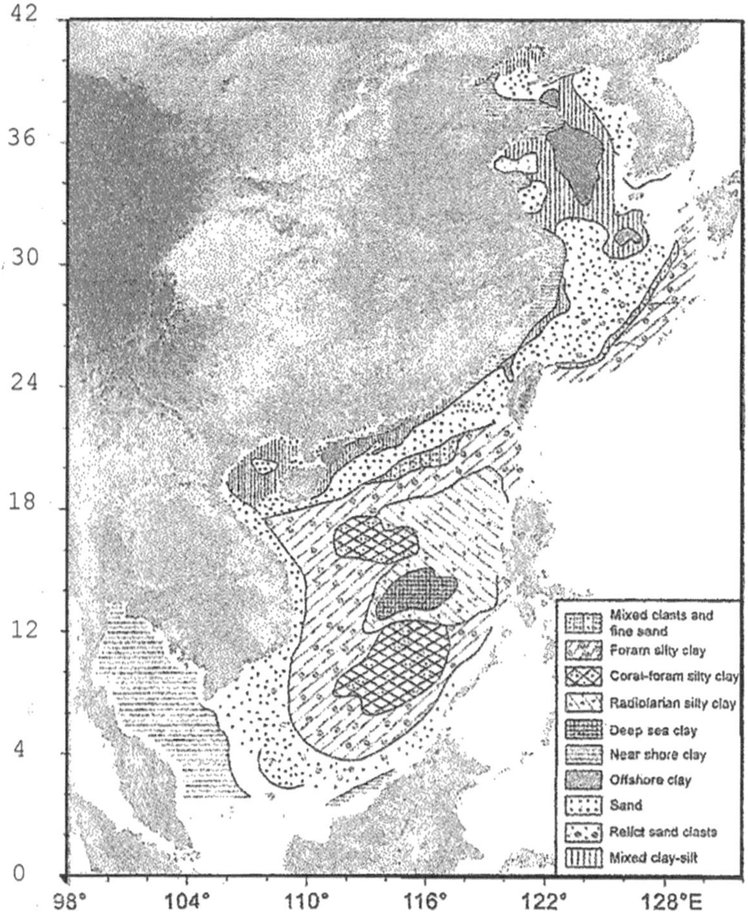

FIGURE 8.2 Sedimentation in the South China Sea (Liu, 1996).

and include diapiric intrusions. The southward continuation of the broad Sunda shelf off Thailand and Malaysia has an arm that curves gradually to the east. This shelf maintains a width of 400 km or more through the SCS.

The clays contributed by the various sources in the SCS area is presented in Figure 8.3 (Liu et al., 2008). These clay minerals are presented in the form of a tripartite graph showing the relative amounts and locations of these clay minerals: kaolinite, smectite (i.e., montmorillinite), illite and chlorite. A review of Figure 8.3 shows that southern clay sediments from Sumatra, Malaysia. Southn Borneo has more kaolinite while Luzon has Smectite. Northern sediments in contrast have more illite and chloride. This difference is related to the differences in volcanism between the different areas.

The SCS shelf along the north-east coast of Borneo narrows to less than 100 km and is still narrower along the Philippine island of Palawan. These narrow shelves are

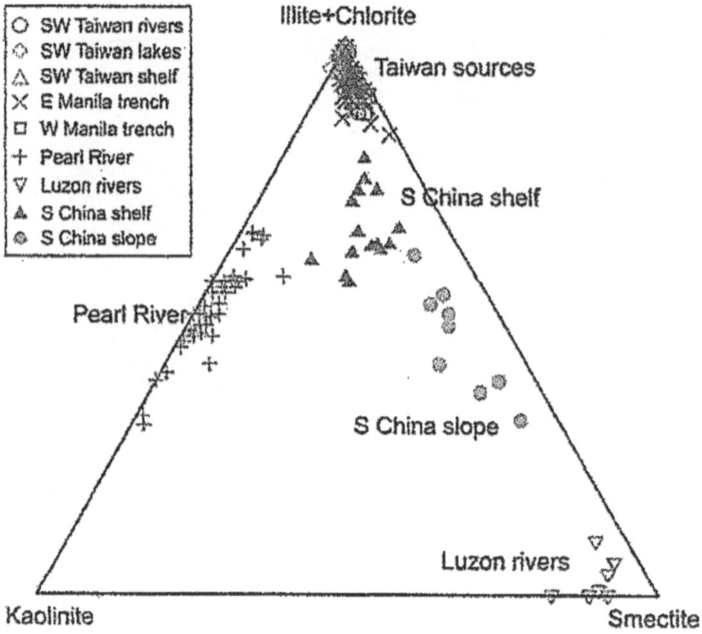

FIGURE 8.3 Tri-partite graph showing the main sources to the China Seas in the ternary diagram of clay minerals (Liu et al., 2008). Reprinted with permission of Elsevier.

partly covered with a mass of shallow coral reefs. In some places, as much as 90 m of sedimentation has occurred during the Holocene.

Around the Philippine Islands, the shelves are narrow, although indentations provide some appreciable shelf widths. The shelves have many coral shoals. On all sides of the Philippines the slopes are unusually steep, averaging 11° down to 2000 m.

8.3 DEEP BASIN SEDIMENTATION

The lower latitude SCS is characterized by intensive mixing in its bottom water. This is due to the movement of North Pacific intermediate and deep water. This movement of water allows for better carbonate preservation.

In the abyssal plain area in the deep basin of the SCS, sediments are predominately composed of 60%–70% clay (illite, montmorillonite, chlorite, kaolinite), and sand that is composed primarily of biogenic debris (mainly radiolarian with diatoms and foraminifera). The mineral content (<1%) of the sediment is mainly quartz, feldspar, and volcanic clastic minerals.

8.4 CHINA SEAS SEDIMENTATION

The SCS deep basin is covered by two main types of sediment: calcareous ooze over the slope and abyssal clay over the central basin below the carbonate compensation depth (CCD). The calcareous ooze covering the slope consists of all biogenic and terrigenous sediments above the CCD. This includes clayey silt, silty clay, and calcareous ooze. The ooze is gray or yellowish gray in color containing both nannofossils and planktonic foraminifera as well as other calcareous skeletons. The carbonate abundance ranges from between 30% and 62% on the upper and middle slope but decreases with increasing water depth due to dissolution, as shown in deposits (Wang, 1999; Wang and Li, 2009, 2014; Wang, Prell and Blum, 2000; Wang et al. 2014).

The abyssal clay consists of a mixture of biogenic and terrigenous material from below the CCD. In the abyssal clay, biogenic skeletons are less than those in shelf and slope sediments. This layer contains a small amount of radiolarian and lesser amounts of diatoms and agglutinated foraminifera. Volcanic substances in sediments are common in the north-eastern region. In addition, coral debris is common in areas surrounding coral reefs or atolls and terranes, especially in the southern region.

8.5 SEDIMENT DISTRIBUTION PATTERNS IN THE CHINA SEAS

Sediment patterns of the China Seas have primarily resulted from a combination of factors: (1) the land–sea interaction between the Pacific Ocean and Eurasia, (2) sea-level fluctuation over glacial periods, and (3) proximity to rivers. The resulting broad shelves were exposed during the last glaciation and their sediments were subsequently washed into deeper waters. During the last postglacial transgression the sediments were redistributed widely within the marginal basins. This resulted in the current sediment distribution characterized by Holocene muds over some parts of the shelf and older sediments on the middle and outer shelf.

REFERENCES

Emery, K.O. and Niino, H. (1963). "Sediments of the Gulf of Thailand and adjacent continental shelf," *Geologic Society of America Bulletin*, 74: 541–554.

Liu, X.Q. (1996). "Sedimentary div. in marginal seas of China," *Marine Geology and Quaternary Geology* 16(3): 1–11 (in Chinese).

Liu, Z.F., Tuo, S.T., Colin, C., Lin, J.T., Huang, C.-Y., Selvara, K. et al. (2008). "Detrital fine grained sediment contribution from Taiwan to the northern SCS and its relation to regional ocean circulation," *Marginal Geology*, 255: 149–155.

Parke, M. L. Jr., Emery, K.O., Szymankeiweiwicz, and Reynolds, L.M. (1971). "Structural framework of continental margin in South China Seas," *American Association of Petroleum Geologists Bulletin*, v55(5): 723–751.

Shepard, F.P. (1973). *Submarine Geology*. New York: Harper and Row, 517pp.

Wang, P.X. (1999). "Response of western Pacific marginal seas to glacial cycles: Paleoceanographic and sedimentological features," *Marine Geology*, 156: 5–39.

Wang, P.X. and Li, Q.Y. (Eds.) [2009] (2014). *The South China Seas Paleoceanography and Sedimentology*. Dordrecht, the Netherlands: Springer, 506pp.

Wang, P., Li, Q., and Li, C.-F. (2014). *Geology of the China Seas*. Burlington: Elsevier, 687pp.

Wang, P.X., Prell, W.L., and Blum, P. (Eds.) (2000). "Leg 184 summary: Exploring the Asian monsoon through drilling in the South China Sea," *Proceedings of Ocean Drilling Program*, Initial reports, 184, College Station, TX.

Part V

Civil Engineering Development

9 Taiwan Strait Transportation (TST) Corridor

9.1 INTRODUCTION

The Taiwan Strait Crossing (TSC) proposed project consists of a combination of both bridges and underwater tunnels with a total length ranging from 150 to 250 km depending on the crossing location. This project will also have influence on many different aspects such as the following: communication between the two sides of the strait, economics, and politics on both sides of the strait. In the following, several issues will be presented. These issues are the following: (1) seafloor characteristics, (2) crossing route, and (3) type of structure(s) to be built to enable crossing the strait.

9.2 PACIFIC OCEAN SEA-FLOOR CHARACTERISTICS

The Pacific Ocean is enclosed by linear mountain chains, trenches, and island systems. This body of water has a surface area of 181×10^6 km^2. These conditions result in deep-sea basins areas in the Pacific Ocean being effectively isolated from the influences of terrigenous sedimentation.

The narrow continental margins of the Pacific Ocean comprise only a small proportion of its total area. The Pacific Ocean has two unique features. The first feature is that it has a number of volcanic islands, located in the central and western parts that include the Taiwan Islands. The second feature is extensive marginal basins of variable sizes that lie close to the margins of the Western Pacific. These features are separated from the deep-ocean basins by trenches or island arcs. The depth of these marginal basins is usually greater than 2 km and they serve as sediment traps. This results in thick layers of sediment in the basins. This type of margin is called a Pacific-type margin. This type of margin has been further subdivided into either the Peru-Chilean-type or Mariana-type. The Peru-Chilean-type is characterized by a narrow shelf with a trench below the slope. In contrast, the island arc, or Mariana-type, exhibits a shallow marginal basin separating the continent from the island arc and trench system as shown in Figure 9.1. The Taiwan Strait sea floor is similar to the Mariana type. Further discussions on sea-floor features are presented by Chaney and Fang (1986) in a state-of-the art paper on marine geotechnology.

In general, sea-floor sediments are predominantly depositional rather than erosional. As a consequence, marine sediments exhibit more uniformity than normally

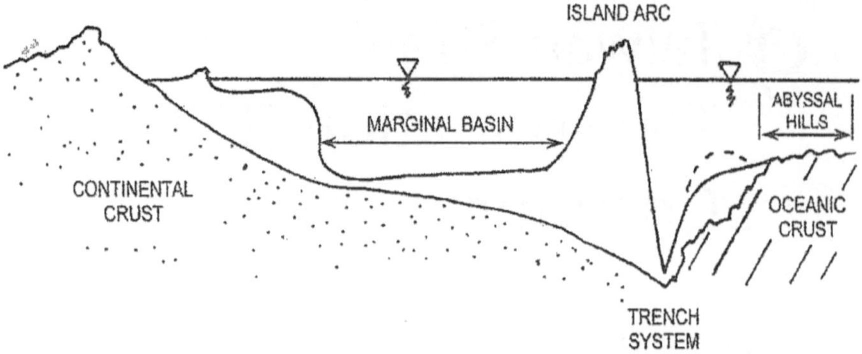

FIGURE 9.1 Mariana type of sea floor at Taiwan Strait (Chaney and Fang, 1986).

found on land. Typically, marine sediments are broadly classified by whether the sediments are land derived (terrigenous) or are the result of marine activity (pelagic). The pelagic sediments can be further divided into inorganic or organic materials. Inorganic pelagic materials are typically clay-size material. In contrast, organic materials are primarily the skeletal remains of marine organisms. These materials are either calcium carbonate ($CaCO_3$) or silica (SiO_2). The presence of calcium carbonate in marine sediments is influenced by biological productivity and the calcium carbonate compensation depth.

9.3 ESTIMATION OF GEOTECHNICAL PARAMETERS

A number of geotechnical soil reports for a variety of projects both on the coastal plain and offshore along the Taiwan Strait have been produced over the last few years as reported by Yin et al. (2003). Based on these site investigations, soils in the Taiwan Strait may be described as three types: clays, silty clays, and silty sand. Estimated geotechnical parameters based on a combination of actual data, correlations, and professional experience have been presented in Table 9.1 by Yin et al. (2003). This estimated soil data can be utilized in parametric studies of possible TSC design scenarios.

9.4 ROUTE SELECTION

Three possible crossing routes (Southern, Middle, and North) from the mainland to the Taiwan Islands are presented in Figure 9.2. These three proposed routes are selected because of the proximity of major cities or metropolitan areas on both Taiwan Island and the mainland. Each proposed route has unique characteristics such as economic, geographic, and environmental conditions. In the following a brief description of these three routes are presented.

A comparison of the alternative proposed TSC routes from the Chinese mainland to Taiwan Island showing the main metropolitan areas is presented in Figure 9.3.

TABLE 9.1
Estimated Geotechnical Model Parameters for Clay, Silty Clay, and Silty Sand

	Clay	Silty Clay	Silty Sand
Basic Properties (ave)			
Specific gravity G_s	2.70	2.65	2.60
Sat. unit weight γ (kN/m^3)	16.5	17.5	19.0
Initial void ratio e_o	1.35	0.80	0.26
Natural water content w (%)	50	30	10
Clay content C (%)	90	30	0
Plastic limit w_p (%)	20	20	—
Liquid limit w_L (%)	60	30	—
Plasticity index I_p (%)	30	10	—
Parameters of Compressibility, Consolidation, and Permeability			
Compression index C_c	0.42	0.15	0.07
Re-compression index C_r	0.055	0.012	0.010
Coef. of secondary consolidation $C_{\nu 0}$	0.011	0.0031	0
OCR (assume NC)	1.0	1.0	—
Coef. of vertical consolidation C_v (m^2/yr.)	0.8	2.0	—
Coef. of hor. consolidation C_h (m^2/yr.)	1.6	4.0	—
Vertical permeability k_v (m/s)	1×10^{-9}	1×10^{-6}	1×10^{-3}
Hor. permeability k_h (m/s)	2×10^{-9}	2×10^{-6}	1×10^{-3}
Strength and Stiffness and Parameters for Undrained Analysis Using Total Stress			
Undrained shear strength C_u/σ_u	0.22	0.15	—
Friction angle ϕ_u (degree)	0	0	—
Undrained Young's modulus (option 1)	70.6	210.2	—
Undrained Young's modulus using $E_u = (500–1000)C_u$ (option 2)	$(500–1000)C_u$	$(500–1000)C_u$	—
Poisson's ratio υ_u	0.499	0.499	—
Strength and Stiffness Parameters for Effective Stress Analysis			
Effective cohesion c' (kPa)	0	0	0
Friction angle ϕ'(deg)	28	32	34
Dilation angle ψ'(deg)	0	0	5
Earth pressure at rest $K_o = 1 - \sin \phi'$	0.53	0.47	0.44
Drained Young's modulus E (option 1) (MPa)	2–20	15–25	5–25
Drained Poisson's ratio v (option 1)	0.25	0.25	0.25
Effective bulk modulus K (option 2)	$98.3p'$	$345.0p'$	$289.8p'$
Effective shear modulus G (option 2)	$59.0p'$	$207.0p'$	$173.9p'$

Source: After Yin et al. (2003). Reprinted with Permission of Taylor & Francis Group.

FIGURE 9.2 Comparison of alternative routes for TSC (Wu and Shen, 2003). Reprinted with permission of Taylor & Francis Group.

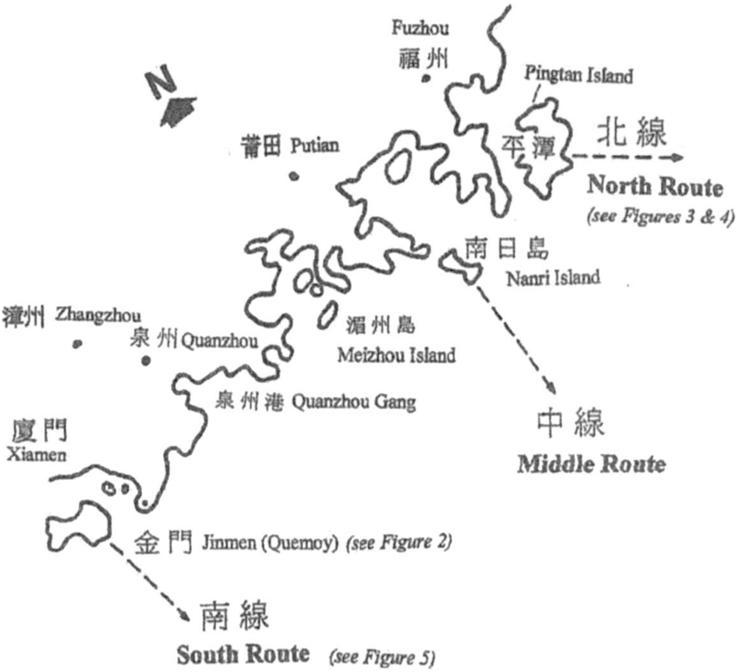

FIGURE 9.3 Proposed Taiwan Strait Crossing (TSC) routes from PRC (Fang, 2003). Reprinted with permission of Taylor & Francis Group.

A review of this figure shows that there are three potential routes under consideration. These routes are the following:

- Southern route—between Jinmen to Penghu Islands-Jiayi (248 km)
  The southern route runs from the islands of Xiamen (i.e., Amoy), Jinmen (Quemoy)in the PRC to Penghu (Pescadores) and from Xingang to Jiayi, then connecting to Tainan and Kaohsiung in the Republic of China (ROC) on Taiwan. Kaohsiung is one of the largest cities and busiest harbors in Taiwan. The major advantage of the southern route is the proximity of several large cities there: Gulagyu, Xiamen, and Jinmen. This location would serve both political and economic purposes for both sides. The proposed bridge sites on the southern TST route are shown in Figure 9.4.
- Middle route between Nanri-Miaoli to Miaoil (147 km)
  The middle route begins from Pinghai, Nanri Dao in the PRC, to Tajia and Taizhong in the ROC (i.e., Taiwan). Taizhong is a large city and metropolitan area in Taiwan. This route on the mainland is relatively undeveloped when compared with the southern and northern routes.

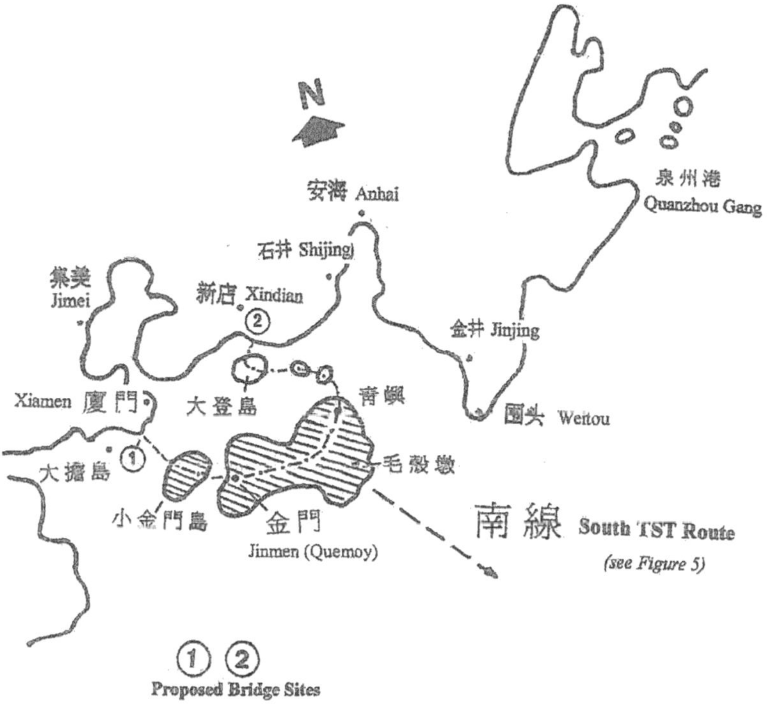

FIGURE 9.4 Proposed bridge sites on the southern TSC route (Fang, 2003). Reprinted with permission of Taylor & Francis Group.

- Northern route (4) between Pingtan–Niu shan–Xinzhu (144 km)
  The north route travels from Pingtan Island, Niu-shan Island in the PRC to Danshui, Taipei and Jilong in the ROC. Taipei is the capital of the Island of Taiwan. Jilong is also a large Taiwanese city with a busy harbor. Figure 9.5 shows the Danshui area, which is an important harbor near Taipei. The proposed bridge sites for the TSC from the north route on the mainland side are shown in Figures 9.6 and 9.6b. Three potential bridge sites (#1, #2, #3) have been proposed. The a1 bridge site is about 3.2 km, the shortest among the three sites. This bridge shown in Figure 9.6b is known as the Pingtan Strait bridge.

FIGURE 9.5 Proposed northern TST route around Danshui area on Taiwan Island (Fang 2003). Reprinted with permission of Taylor & Francis Group.

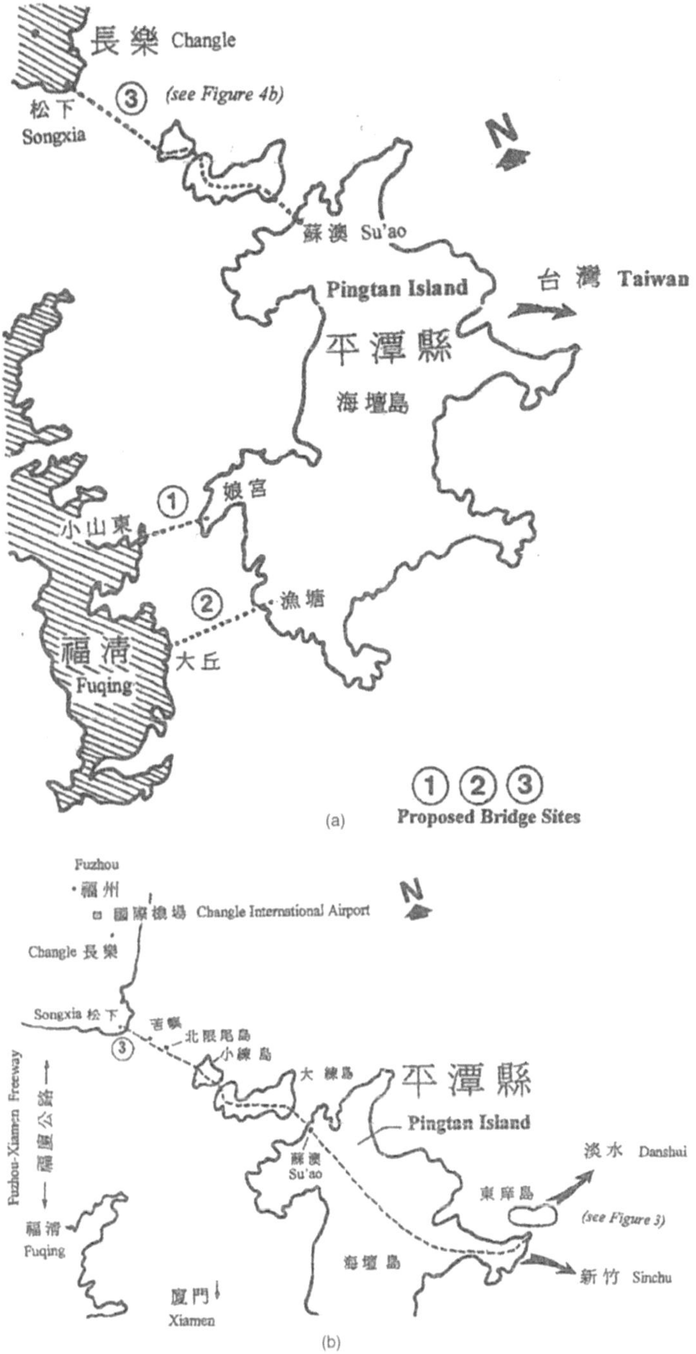

FIGURE 9.6 Proposed bridge sites on the north TST route from mainland side: (a) possible locations of the three bridge sites, and (b) detail of one (#3) of the proposed bridge sites (Fang, 2003). Reprinted with permission of Taylor & Francis Group.

9.5 TYPE OF CONSTRUCTION

The TSC will probably involve a combination of bridges and undersea tunnel. A number of issues will be involved in the design process. These issues are in part: design concept, construction technology, availability of materials, skilled laborers, equipment availability, transportation networks (Wu and Shen, 2003). In the following the issue of design concepts for both a tunnel and bridge will only be explored. The remaining issues have been discussed in considerable detail in a special issue of the *Journal of Marine Georesources and Geotechnology* (Volume 21, Number 3–4, 2003). The special issue was entitled Preliminary Study of the Taiwan Straiat Tunnel Project. The editors for this special issue were Professors Hsai-Yang Fang, Funan Peng, and Zhiming Wu.

9.5.1 Tunnel

An idealized longitudinal cross section of the TSC for the northern route is presented in Figure 9.7.

The selection of the transportation corridor route depends on two primary factors: (1) nearness to the population centers at both ends, and (2) shortest combined tunnel and bridge lengths.

Two basic design approaches are available. These two approaches involve bridges connecting Pingtan and Niu-shan Islands to the mainland. This is followed by either a tunnel located below the seabed (Figure 9.8) or a seabed structure (Figure 9.9). The subseabed tunnel structure would involve the excavation of a series of working shafts at the Taiwan and Niu-shan Islands to minimize the construction time, ventilation, and provide an escape route for emergencies. In addition, an artificial island in mid-strait with two working shafts is proposed. A schematic illustration of this arrangement is presented in Figure 9.7. The artificial island is proposed to be 65 m high, which gives it approximately a 5m freeboard above the highest expected sea level. The dimensions of this artificial island are planned to be 2000 m^2 (100 m × 200 m) at the top. The corresponding base is planned to be 101,764 m^2 or approximately 273 m × 373 m giving a side slope of 1:1.33. This gives an approximate volume of

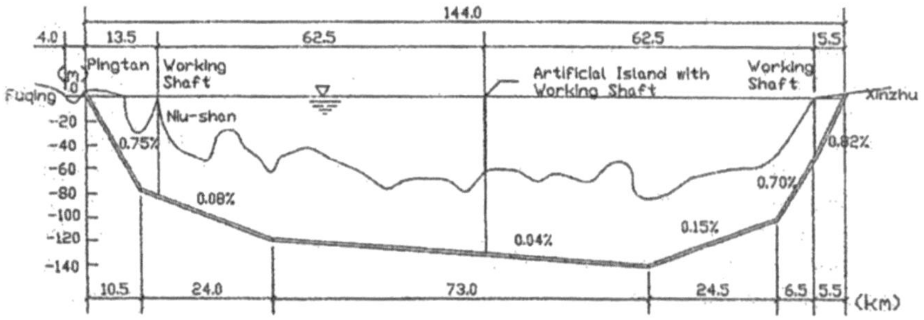

FIGURE 9.7 Longitudinal cross section of proposed TSC (Wu and Shen, 2003). Reprinted with permission of Taylor & Francis Group.

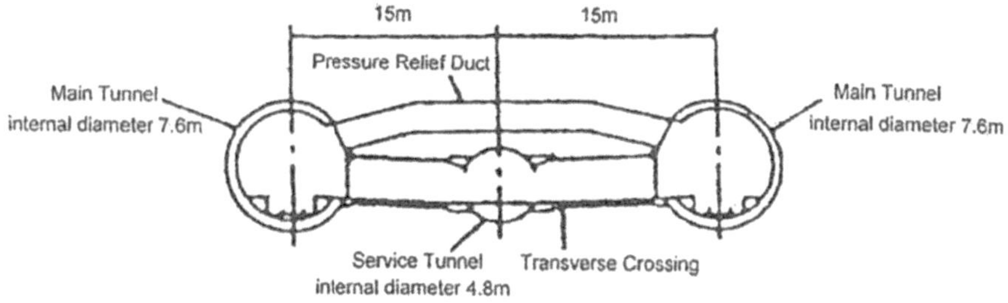

FIGURE 9.8 Cross section of sub-seabed casing (Wu and Shen, 2003). Reprinted with permission of Taylor & Francis Group.

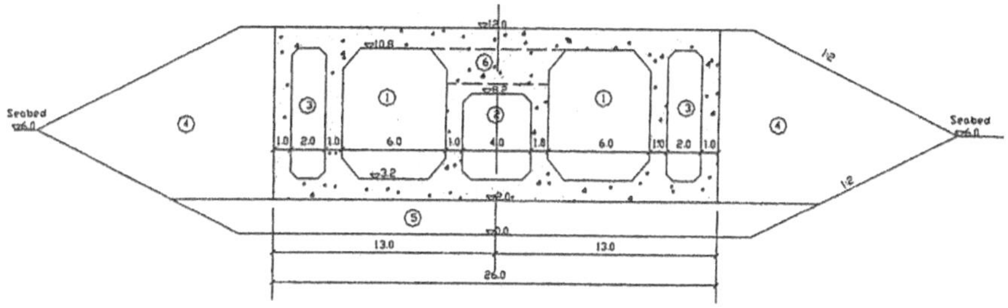

FIGURE 9.9 Typical cross section of a submerged seabed casing: (1) Main operation crossing, (2) Auxiliary crossing, (3) Compartment, (4) Stone or concrete block protection piles, (5) Gravel cushion, and (6) Air pressure relief duct (Wu and Shen, 2003). Reprinted with permission of Taylor & Francis Group.

$3.6 \times 10\ m^3$. Two reinforced concrete working shafts with an internal diameter of 15 m will be installed in the island to connect with the undersea tunnel.

9.5.2 Bridge

9.5.2.1 Conceptual Design

The first step in any bridge design is to visualize the structure in order to determine its primary function and imagined performance. This step occurs before any analysis or detail design can occur. The design process includes the following steps: selection of bridge system, materials, proportions, dimensions, foundations, aesthetics, and how it will fit into the surrounding landscape and environment.

9.5.2.2 Bridge Types

The type of bridge to be employed in a given application is normally dictated by a number of factors such as the following: design loads, surrounding geographical features, geology, passing land, clearance requirements below the bridge, width requirements, available construction materials, construction cost, and construction time required (Duan and Chen, 2003). In Table 9.2 a comparison between types of bridges and span length is presented.

TABLE 9.2
Types of Bridges and Applicable Span Length

Bridge Type	Span Range (m)	Leading Bridge and Span Length
Prestressed concrete girder	10–300	Stolmasundet, Norway, 301 m
Steel I/Box girder	15–376	Sfalassa Bridge, Italy 376 m
Steel truss	40–550	Quebec, Canada. 549 m
Steel arch	50–550	Shanghai Lupu, China. 550 m
Concrete arch	40–425	Wanxian, China. 425 m (steel tube filled concrete)
Cable-stayed	110–1100	Sutong, China. 1088 m
Suspension	150–2000	Akaski-Kaikyo, Japan 1991 m

Source: Duan and Chen (2003). Reprinted with permission of Taylor & Francis Group.

9.5.2.3 Aesthetics—Harmonizing with Surroundings

A bridge must through its structure and form blend with and harmonize with its surroundings. These requirements are in addition to meeting both its structural load requirements and perform as a thoroughfare. In the following is a list of general aesthetic requirements that need to be considered (Table 9.3).

9.5.2.4 Loading

(1) Seismic Design

The seismic criteria for a major bridge are developed based on two levels of earthquake loading. These levels are (1) safety evaluation earthquake (SEE), and (2)

TABLE 9.3
Aesthetics Requirement

- "Choice of a clean and simple structural system," like a beam, a frame, an arch or a suspended structure. The bridge must look trustworthy and stable.
- "Good proportion in all three dimensions" between the structural members or between length and depth of bridges openings.
- "Good order of all the lines of edges of a structures" that determine the appearance. One should limit the number of directions that cause unrest, confusion, and worried feelings. For the transition from a straight to a curved lines the curvature should steadily increase like a second order parabola.
- The compatible integration of a structure into its environment, into the landscape of city. This is particularly important with regard to the scale of the structure compared to the scale of the surroundings
- The choice of the materials has considerable influence on the aesthetic effects.
- Simplicity and restriction to the pure structural shape is important.
- Pleasing appearance can be enhanced by color,
- The space above the bridge should be shaped in such a way that the driver experiences the bridge and gets a comfortable feeling
- A structure must be designed so that the flow of forces is evident to the casual observer.
- Moderate aesthetic lighting can enhance the appearance of a bridge at night.

Source: Svensson (1998). Reprinted with permission of Taylor and Francis Group.

functional evaluation earthquake (FEE) (CalTrans, 2019). CalTrans further divides the bridge categories into three parts. These parts are the following (Table 9.4):

- Important Bridges
  Bridges that provide vital links for the area
- Recovery Bridge
  Bridges that are important for the recovery of the area
- Ordinary Bridges
  Bridges that do not fit into the previous two categories.

In the following is a list of general seismic resistance guidelines that need to be considered.

TABLE 9.4
Seismic Resistance Guidelines

- Bridge type, component and member dimensions, and aesthetics shall be investigated to reduce the seismic demands to the greatest extent possible. Aesthetics should not be the primary reason for producing undesirable frame and component geometry.
- Bridges should be as straight as possible. Horizontally curved bridges complicate and potentially magnify seismic responses.
- Superstructures should be continuous with as few joints as possible. Necessary restrainers and sufficient seat width shall be provided between adjacent frames at all expansion joints, and at the seat-type abutments to eliminate the possibility of unseating during a seismic event. Simply supported spans should not rely on abutments for any seismic resistance.
- Skew angles should be as small as possible. That is, abutments and piers should be oriented as close to perpendicular to the bridges longitudinal axis as possible even at the expense of increasing the bridge length. Skewed abutments and piers are highly vulnerable to damage due to undesired rotation response and increased seismic displacement demands.
- Adjacent frames or piers shall be proportioned to minimize the differences in the fundamental periods and skew angles, and to avoid drastic changes in stiffness and strength in both the longitudinal and transverse directions. Dramatic changes in stiffness result in damage to the stiffer frames or piers. It is strongly recommended (CalTrans 2001) that the effective stiffness between any two bents within a frame, or between any two columns within a bent, do not vary by a factor of more than two. Similarly, it is highly recommended that the ratio of the shorter fundamental period to the longer fundamental period for the adjacent frame in the longitudinal and transverse directions is larger than 0.7. Each frame shall provide a well-defined load path with predetermined plastic hinge locations and utilize redundancy whenever possible. Balanced mass and stiffness distribution in a frame results in a structure response that is more predictable and is more likely to respond in its fundamental mode of vibration. Simple analysis tools can then be used to predict the structure's response with relative accuracy, whereas irregularities in geometry increase the likelihood of a complex nonlinear response that is difficult to predict accurately by elastic modeling or plane frame inelastic static analysis.
- Seismic protective devices, that is, energy dissipation and isolation devices, may be provided at appropriate locations, thereby reducing the seismic force effects. The energy dissipation devices are intended to increase the effective damping of the structure by adding dampers to the structure thereby reducing forces, deflections, and impact loads. Isolation devices are used to lengthen the fundamental mode of vibration by providing isolations at bearing locations so that the structure is subject to lower earthquake forces.

(Continued)

TABLE 9.4 (*Continued*)

- For concrete bridges, structural components shall be proportioned to direct inelastic damage into the columns, pier walls, and abutments. The superstructure shall have sufficient over strength to remain essentially elastic if the columns/piers reach their most probable plastic moment capacity. The superstructure-to-substructure connection for non-integral caps may be designed to fuse prior to generating inelastic response in the superstructure. The girders, bent caps, and columns shall be proportioned to minimize joint stresses. Concrete columns shall be well proportioned, moderately reinforced, and easily constructed. Moment resisting connections shall have sufficient joint shear capacity to transfer the maximum plastic moments and shears without joint distress.
- Initial sizing of columns should be based on slenderness ratios, bent cap depth, compressive dead-to-live load ratio, and service loads. Columns shall demonstrate dependable post yield displacement capacity without an appreciable loss of strength. Thrust–moment–curvature relationships should be used to optimize a column's performance under service and seismic loads. Abrupt changes in the cross section and the capacity of columns shall be avoided. Columns must have sufficient rotation capacity to achieve the target displacement ductility requirements.
- For steel bridges, structural components shall be generally designed to ensure that inelastic deformation occur only in the specially detailed ductile sub-structure elements. Inelastic behavior in the form of controlled damage may be permitted in some of the superstructure components such as the cross frames, end diaphragms, shear keys, and bearings. The inertial forces generated by the deck must be transferred to the substructure through girders, trusses, cross frames, lateral bracings, end diaphragms, shear keys, and bearings. As an alternative, specially designed ductile end-diaphragms may be used as structural mechanism fuses to prevent damage in other parts of structures.
- Steel multicolumn bents or towers shall be designed as ductile Moments-Resisting Frames (MRF) or ductile braced frames such as Concentrically Braced Frames (CBF) and Eccentrically Braced Frames (EBF). For components expected to behave inelastically, elastic buckling (local compression and shear, global flexural, and lateral torsion) and fracture failure modes shall be avoided. All connections and joints shall preferably be designed to remain essentially elastic. For MRFs, the primary inelastic deformation shall preferably be columns. For CBFs, diagonal members shall be designed to yield when members are in tension and to buckle inelastically when they are in compression. For EBFs, a short beam segment designated as a *link* shall be well designed and detailed.
- The ATC/MCEER recommended LRFD *Guidelines* (2001) classify the earthquake resisting systems (ERS) into permissible and not recommended categories based on consideration of the most desirable seismic performance, ensuring wherever possible post-earthquake serviceability.

Source: Duan and Chen (2003). Reprinted with permission of Taylor & Francis Group.

(2) Wind Design

The importance of the aerodynamics of bridge structures was demonstrated by the 1940 collapse of the long span Tacoma Narrows bridge (Ammann et al. 1941; Tacoma, 2003). The objectives of wind design for long span bridges are presented in Table 9.5.

To achieve the above objectives careful consideration of the following have to be made: structural forms, stiffness, cross section shape, details, and damping (Cai and Montens 2000; Kubo 2004). The above issues need to be addressed to improve the aerodynamic behavior of bridges and to reduce wind effects. Methods to improve aerodynamic behavior of bridges are presented in Table 9.6.

TABLE 9.5
Objectives of Wind Design for Bridges

- Provide necessary strength to resist wind-induced static blow/drag forces.
- Ensure the critical wind velocity has a very low probability of occurrence to avoid flutter effects, a self-excited unstable condition.
- Control magnitude of buffeting response to reduce influences on fatigue of the bridge and users' comfort.

Source: Duan and Chen (2003). Reprinted with Permission of Taylor & Francis Group.

TABLE 9.6
Elements for Improving Aerodynamic Performance of Bridge Design

- *Structural form:* Suspension bridges, cable-stayed bridges, arch bridges, and truss bridges, due to the increase of rigidity in this order, have generally aerodynamic behaviors from the worst to the best. A truss-stiffened super-structure blocks less wind, and is more favorable than a girder-stiffened one. But a truss-stiffened bridge is generally less stiff in torsion.
- *Stiffness:* For long span bridges, it is not economical to add more material to increase the stiffness. However, changing the structural shapes and boundary conditions, such as deck and tower connections in cable-stayed bridges, may significantly improve the stiffness. Cable-stayed bridges with "A" or inverted "Y" shape towers have higher torsional frequency than the bridges of "H" shape towers.
- *Cross section shape and its details:* A streamlined section blocks less wind thus has better aerodynamic behavior than a bluff section. Small changes in section details may significantly affect the aerodynamic behavior.
- *Damping:* Concrete bridges have higher damping ratios than steel bridges. Consequently, steel bridges have more wind-induced problems than concrete bridges. Protective damping systems may be employed to reduce aerodynamic vibration.

Source: Duan and Chen (2003). Reprinted with permission of Taylor & Francis Group.

(3) Vessel Collision Design

A major turning point in the development of vessel collusion design criteria occurred in thc 1980 collapse of the Sunshine Skyway Bridge. This bridge was located crossing Tampa Bay in Florida. The bridge collapsed because of the collision of an empty bulk container ship with one of bridge's anchor piers. This collision resulted in 396 m of the southbound span collapsing killing 35 individuals when their vehicles fell into the bay. After this accident, a series of guidelines were developed (Knott and Larsen, 1990). Table 9.7 presents a set of recent vessel collision guidelines by Knott and Prucz (2000).

TABLE 9.7
Vessel Collision Guidelines

- Bridges should be located away from turns in the channel. The distance to the bridge should be such that vessels can line up before passing the bridge, usually at least eight times the length of the vessel. An even larger distance is preferable when high currents and winds are likely to occur at the site.
- Bridges should be designed to cross the navigation channel at right angles and should be symmetrical with respect to the channel.
- An adequate distance should exist between bridge locations and areas with congested navigation, port facilities, vessel berthing maneuvers, or other navigation problems.
- Locations where the waterway is shallow or narrow, so that bridge piers could be located out of vessel reach, are preferable.
- The vertical clearance provided in the navigation span should be based on the highest vessel that uses the waterway in a ballasted condition and during periods of high water level. The vertical clearance requirements need to consider site-specific data on actual and projected vessels.
- A clear load path from the location of the vessel impact to the bridge foundation needs to be established and the components and connections within the load path must be adequately designed and detailed.
- The number of approach piers exposed to vessel collision should be minimized, and horizontal and vertical clearance considerations should also be applied to the approach spans.
- Bridge protection alternatives such as fender systems, dolphin protection systems, island protection systems and floating protection systems should be carefully evaluated to develop a cost-effective a solution.

Source: Knott and Prucz (2000). Reprinted with Permission of Taylor & Francis Group.

9.6 SUSTAINABLE CONSTRUCTION

The goal of sustainable construction is based on finding a balance solution between economic development and protection of the environment (i.e., natural resource preservation) (Chen and Chen, 2003). This goal leads to the following principles:

- Social progress must fit the needs of the public
- Pursue economic development to promote the quality of life
- Protect environment to preserve natural environment
- Utilize resources efficiently to achieve sustainable utilization of finite resources that we have.

REFERENCES

Ammann, O.H., Theodore, V.K., and Woodruff, G.B. (1941). *The Failure of the Tacoma Narrows Bridge*, Federal Works Agency, Washington, DC

Cai, C.S., and Montens, S. (2000). "Wind effects on long-span bridges," in W.F. Chen and L. Duan (Eds.), *Bridge Engineering Handbook*, CRC Press, Boca Raton, FL.

CalTrans, California State Department of Transportation (2019). Seismic design criteria. Version 2.0, April, State of California, Sacramento.

Chaney, R.C., and Fang, H.Y. (1986). "Static and dynamic properties of marine sediments," *Proceedings of Symposium on Marine Geotechnology and Nearshore/Offshore Structures*, Shanghai, China, ASTM STP 923, pp. 74–111.

Chen, R.H., and Chen, K.H. (2003). "Control and recovery of environment after the construction," *Marine Georesources and Geotechnology*, 21: 237–247.

Duan, L., and Chen, W-F. (2003). "Basic design criteria for bridges crossing open sea and bay area," *Marine Georesources and Geotechnology*, 21: 289–305.

Fang, H-Y. (2003). "The ocean community," *Marine Georesources and Geotechnology*, 21: 155–166.

Knott, M.A., and Larsen, O.D. (1990) Guide specification and commentary vessel collision design of highway bridges. Report No. FHWA-RD-91-006. Washington, DC: US Department of Transportation, Federal Highway Administration.

Knott M., and Prucz, Z. (2000). *Vessel Collision Design of Bridges: Bridge Engineering Handbook*. CRC Press LLC, Boca Raton, FL.

Kubo, Y. (2004). "Structure configurations based on wind engineering," in W.F. Chen, and E.M. Lui (Eds.), *Structural Engineering Handbook*, 2nd ed., CRC Press, Boca Raton, FL.

Svensson, H.S. (1998). "Aesthetics of steel bridges," in *The Proceedings of 1998 World Steel Bridge Symposium*, Chicago, IL.

Tacoma. (2003). Internet references: www.civ.toronto.edu/funstuff/disaster/tacoma.html; www.lib.washington.edu/specialcol/tnb/; www.civeng.carleton.ca/Exhibits/Tacoma_Narrows/; www.enm.bris.ac.uk/research/nonlinear/tacoma/tacoma.html.

Wu, Z., and Shen, Z. (2003). "Design concept and engineering techniques of Taiwan Strait Crossing," *Marine Georesources and Geotechnology*, 21: 269–278.

Yin, J.H., Liao, H.H., Zhou, C., and Cheng, C.M. (2003). "Estimation of marine soil parameters for preliminary analysis of geotechnical structures in the Taiwan Strait Connection Project," *Marine Georesources and Geotechnology*, 21: 167–182.

Index

*Bolded locators indicate tables.
*Italicized locators indicate figures.

A

Airy theory, 57–58, 203–206
Almagor, Gideon, 91, 125
Anderson, Donald G., 91, 178, 180, 185, 197, 209
AnhDan, L., 162
Ansal, A.M., 186
AOCR, 185
apparent over-consolidation hypothesis, *see* AOCR
atmospheric circulation, 36, *37*, 38, *39*, 43
Atterberg limits, 81, 85, 87, 94, 142
Attwooll, W.J., 95
Augustesen, A., 161

B

back-arc basin, 12, 16
Bennett R.H., 75, 146
Benthos, 154
Berger W.H., 82
Bhatia, S.K., 200
bioturbation, 151–155
  bioturbation index (BI), 152, **155**
  deep-sea sediment, bioturbated, 154–155
  shallow water, 154
Booker, J.R., 207
Booth, J.S., 142, 146
Brown, J.D., 146, 198
bulk solids, 114
Busch, W.H., 143

C

calcareous materials
  compressibility, 119–125
  consistency, 116–119
  deformation parameters, 131–133
  density, 112–116
  permeability, 125
  shear strength, 126–128
  water content, 112–116
calcareous ooze, *121*, *122*, 220–221
calcium carbonate material, description, **87**
capillary waves, 53–54
carbonate compensation depth, 220, 226
Caroline Plate, 13
Casagrande preconsolidation pressure, 98
Castro, G., 209
CCD, 220–221
Chaney, R.C., 91, 176, 178, 200, 202
Chingchang, B., 24
Christensen, R.W., 164–165
Christian, J.T., 209
circulation cells, 38
circulation pattern, 38, *39*, 44, 48
clays, 92–111
  behavior, 181–198
  compressibility, 96–98
  consistency, 94–96
  density, 94
  Drammen clay, 98, 183, 193, **194**
  isotach behavior, *162*
  minerals, 72, 77, 88, 95, 173–175, 219, *220*
  particles, 73–76, 95, 102
  permeability, 99–105
  pore pressure build-up, 183–198
  predicting cyclic response, 198
  shear strength, 105–112
  strength determination, 182–183
  water content, 94
CLRL, 182
CLS, 168–170
colloids, 73
conducted cyclic triaxial tests, *see* CTX test
consolidation
  normal consolidation, 95, 105, *106*, 115, 123, 163, **227**
  over consolidation, 146, *149*, 181–185, 188, 190, 194, 197, 209
  primary, 119–123
  secondary, 123–155
  under-consolidation, 95, 105–106, 209
constant void ratio, *see* CVR
construction, sustainability, 238
Coriolis effect, *38*, 43–45, 47
creep
  behavior, 161, 167–168
  characteristics, 163
  conditions, 164
  properties, 163
  rate, 162–164
  time, 96
Cretaceous, 14, 16, 20
critical level of repeated loading, *see* CLRL
critical state line, *see* CSL

Cromwell Current, 49
cross-section, 14, 91
crustal thickness, 16
CSL, 168, 186
CTX test, 173, 176, *191*
Cuny, R.W., 195
currents
  oceans, water exchange, 50
  surface wind-driven, 43–49
  types, *44*
  vertical currents, density driven, 49–50
CVR, 168, 202
cyclic loading
  clay minerals, 173–175
  post-cyclic loading behavior, 209–210
  sands, 175
  theoretical background, 168–170
  threshold strain concept, 170–173
cyclic state line, *see* CLS
cyclic stress ratio, 183–184, 186, 190, *192*, 199–200; *see also* CLS

D

Dahl, A.G., 142, 146
deep-sea drilling project, *see* DSDP
Demars, K.R., 91, 113, 128
Design Water Levels, *see* DWL
Di Benedetto, H., 162
diatom ooze, 134, 136, **137**, **138**
diatoms, 133–134, 220–221
dispersion, 56–57, 73, 75, 139
Dobry, R., 186, 189
double-layer interactions, 75
down welling, 48
Drnevich, V.P., 188–189
DSDP, 82
DWL, 43
dynamic loading
  theoretical background, 168–170
  threshold strain concept, 170–173

E

earth–moon–sun system, 40
earth system processes, 33–36
earthquakes, 21, 25, *27*, *28*, 53, 61–64
East China Sea, 23–24, 217
East Sub-basin, *see* ESB
EEZ, 5, 7
Egan, J.A., 185
Ekman spiral, 44, 46
Ekman transport, 46–48
electromotive force, *see emf*
electro-neutrality, 73
emf, 33
equations, 43, 55–60, 63, 65, 168, 185, 200, 205–207
Equatorial Counter-current, 49
Erken, A., 186
ESB, 14
Eurasian Plate, 12–13, 23–25

F

Fang, H.Y., 176, 202, 225
FEE, 235
Finn, W.D.L., 200
fine-grained sediments, 94–111
  compressibility, 96–98
  consistency, 94–96
  density, 94
  permeability, 99–105
  shear strength, 105–111
  water content, 94
flocculation, 73, 75
floccules, 75
flocs, 75
Focht, J.A., 209
foram ooze, 125, 128, *130*
foram solids, 114
France, J.W., 209
Fry, Z.B., 195
Fujian Province, 21, 23, 25
Fukueicuiao Cape, 23
functional evaluation earthquake, *see* FEE

G

Gagua Ridge, 23
Gardner, W.S., 209
geography, 3–4
Gradstein, F., 17
granular materials, behavior, 198–203
  liquefaction phenomena, 200–203
gravity wave generation
  Airy wave theory, 57–58
  cnoidal wave theory, 60–61
  linear wave theory, 54–58
  Stokes wave theory, 58–59
Guangdong Province, 23

H

Hadley cells, 38
Hardin, B.O., 188–189
Hendrickson, J., 58–59
Herrmann, H.G., 91, 180
high plasticity (MH), 119
hydration, 73, 75

I

Idriss, I.M., 198
Infrared band, 34–35
International Ocean Drilling Program, *see* IODP
IODP, 16
Ionization, 73

J

Jardine, R., 162
JOIDES, *184*
Jurassic, 20

K

Keller, G.H., 143, 146
Kiyota, T., 162
Kuroshio Current, 48–49, 51, 218

L

Lade, P.V., 161–2
Lancaster, J.K., 147
land–sea interaction, 221
Law for the Sea, **4**, **5**, 6
LeBoeuf D., 197
Lee, H.J., 114, 126–127
Lee, K.L., 209
Lefebvre, G., 197
Liingaard, M., 161
limestones, 82, 85
 term, 88
linear, 57–58, 203–206
liquefaction, 200–203
 implications, 200–202
 mechanism, 200–202
 variations, 202–203
liquid limit, 81, 91, 92, 94, *96*, 116, *117*, 142, *145*, 146, **148**, **227**
Lisitzin, A.P., 82
loading
 cyclic, 168–175
 dynamic, 168–175
 seismic, 168
 sustained, 161–168
 wave, 174, *202*
long valley, 23

M

Macclesfield Bank, 4, 17–18
MacFarlane I.C., 146
Macky, T.A., 197
marginal basins, 18, 20, 217, 221, 225
 characteristics, **15**
 marginal ocean basin, 14, 16, 50
marine carbonate sediments
 engineering properties, 85
 sediments, summary, *86*
marine sediments
 apparent overconsolidation behavior, **100–101**
 calcareous materials, 112–133
 characteristics, 91
 characterization, 69–88
 classification, *80*
 clays, 92–111
 compressibility, 91–155
 distribution, **70**, *71*
 fine-grained sediments, 92–111
 index, 91–155
 organic sediments, 138–155
 physical properties, 91
 siliceous sediments, 133–138
 strength, 91–155
 under loading, 161–175
marine soil, 69, 94
Martin, G.R., 200
Masing criterion, 187, 193
Matsushita, M., 162
Mesozoic, 14, 16
Mesri, G., 123–124
Mitchell, J.K., 103, 161, 165–166
Murayama, S., 164

N

Nan-Ao Island, 23
nanno solids, 114
nannoforam ooze, 114
nannofossil, 85, **100**, **104**, 105, 112, **113**, 114, *119*, 124, 221
nannofossil ooze, *119*
nano ooze, 104, 124
NC, 95, 105, *106*, 115, 123, 163, **227**
Neogene, 23
NG, 197
normal consolidation, 95, **105**, 105–106, 120, 185–186, **196**, 197; *see also* NC
North and South Equatorial Currents, 49
North Pacific gyre, 46, 49
North Palawan Island, 17–18
Northwest Sub-basin, 17–18
NWSB, 14

O

ocean currents
 types, *44*
 westward intensification, 47
oceanographic factors
 currents, 43–50
 earth system processes, 33–36
 tides, 40–43
 winds and storms, 36–39
ocean surface waves, 53–54
OCR, 146, *149*, 181–185, 188, 190, 194, 197, 209
Odell, K.T., 143
Oluanpi Cape, 23
ooze
 calcareous ooze, *121*, *122*, 220–221
 diatom ooze, 134, 136, **137**, **138**
 foram ooze, 125, 128, *130*
 nannoforam ooze, 114
 nannofossil ooze, *119*
 radiolarian ooze, **83**, 134, 136, **137**
 siliceous ooze, 133

organic sediments, 138–155
bioturbation, 151–155
compaction, 146–151
compressibility, 146–151
consistency, 142–146
density, 141–142
gel, 139
marine clays (South East Asia), 139
organic material (South East Asia), 139
permeability, 151
water content, 141–142
Osgood, W.T., 188
over-consolidation ratio, *see* OCR
ozone, 33, 35

P

Pacific Ocean
sea floor, characteristics, 225–226
Palawan Trough, 4
Paleogene, 14, 23
Paleozoic, 14
Pamukcu, S., 187, 197
Paracel Islands, **5**, 17, 18
Parke, M.L., 218
Peikang Basement High, 25
Peng, F., 24
People's Republic of China, *see* PRC
Philippine Boundary Belt, 11
Philippine Mobile Belt, 11, **12**, 13
Philippines, 3–4, **5**, 7, 13, **21**, 48–49, 220
PI, 92, 95, 107, 142, **164**, 173, 182, 185, 188–189, 193, **196**, 197
Pingtan Island, 23, 230
plasticity index, 81, 91–92, 94, 96, 106–107, 110, 116, *117*, 123, 142, *145*, 146, **148**, 163, 173, **196**, 227; *see also* PI
plates, 11–12, 14, 16, 19–20
PMB, 13–14; *see also* Philippine Mobile Belt
political divisions, 4–8
pore pressure
build-up, 183–187
excess, 106, 128, 152, 161, 170, 172–173, 181–182, 184–185, 209
residual, 183–184, 200
water pressure–strain curve, 174
Potts, D.M., 184, 209
Praeger, S.R., 209
PRC, **4–5**, 6–7, *228*, *229*, *230*
Pusch, R., 139, 142–143

R

radiolarian ooze, **83**, 134, 136, **137**
Rahman, M.S., 207
Ramberg, W., 188, 190, 198
Ramberg-Osgood model, 190, *191*
Rashid, M.A., 146
RC test, 173, 176
Reed Bank, 4, 17–18
Republic of China, *see* ROC
resonant column test, *see* RC test
resources, 5–8
Richards, A.F., 151, 206
Richter magnitude, 62
ROC, 4, 20, **21**, 229–230
rocks, "flow", 18
Rutka, A., 146
Ryukyu Arc, 23
Ryukyu Islands, 23
Ryukyu Trench, 23

S

Saada, A.S., 197
safety evaluation earthquake, *see* SEE
salinity and temperature, *see* SST
Sangrey, D.A., 173–174, 182, 185, 209
Scarborough Seamount, 4
SCS, 3–7, 11, 16, 18–19, 21, 23, 28, 37, 48, *78*, 217–220
basin, 14
circulation patterns, *39*
throughflow, 50
political events, **4**, **5**
sea areas, 5, *6*
seabed, 7, 48, 71, 112, 136, 151, 176
deformable, 207–209
rigid, 206–207
sediments, 168, 174, 203
wave interactions, 203–209
seafloor spreading, 14, 16–19, 21
SEASWAB, 207
seascape, 70
sea state, 61, **62**
secondary compression, 123–125, 161, 163
sediments
basic division, *72*
characteristics, 91–92
characterization, 72–79
classification, 79–82
deep-sea, **83**, **82**, **103**
engineering properties, **92**
fine-grained, 94–111
organic, 138–155
plasticity characteristics, 142
sediment thickness, 16, **101**
sedimentary environment, 71, 79, 81
sedimentary rocks, 23, 69
sedimentation, 217–221
siliceous, 133–151
structure disintegration, 77–79
term, 69
SEE, 234
Seed, H.B., 180, 207, 209–210
seismic loading, 168
SHANSEP method, 193
shear strength, 91–92
profiles, 105, 107, *108*, *109*

undrained, *107*, 173, 181–183, 185, 209–210, **227**
Shibata, T., 164–165
siliceous ooze, 133
siliceous sediments, 133–151
compressibility, 136–138
consistency, 136
density, 136
permeability, 136
shear strength, 138
water content, 136
silts
behavior, 181–198
strength determination, 182–183
Singapore, 3
Singh, A., 163, 165–166
Singh. R.D., 185, 209
Skjelbreia, L., 58–59
soil
calcareous, 87, 119, *120*, 122, 126, *132*
carbonate, 87, 118–119
classification, 71, 79, 81
compaction, 146–147
grains, 99, 125, 127, 136, 201
matrix, 126
organic, classification system, **140**
parameter modeling, 175–181
particles, 76–79
pores, 99, 125, 138
structure, 123, 161, 166, 199, 201
term, 69
South China Sea
opening, 16–23, **21**
river discharge, *22*, **22**
topographic features, 14–16; *see also* SCS
Southwest Sub-basin, *see* SWSB
Spratly Islands, 4, **5**, 7, 20, **21**
SS, 168–170, 175, 193
SSCR, 163, **164**
SST, 50
steady state, *see* SS
steady-state creep rate, *see* SSCR
Stokes, G.G., 58
Stokes wave theory, 54, 58, 60
storm surge, 64–65
gust velocity, 65
winds, fastest miles, 64–65
Strait of Malacca, 3
stress level, 119, 131, 133, 162, 169, 181–182, 209
stress–strain behavior, 174, 187, 198
stress–strain curve, 161, 183
Suhayda, J.N., 187, 207–208
Sunda Plate, 11–12
Sunda Shelf, 16, 21, 217–219
surface-held water, 73
surface wave, 53–54, 56, 59, 206–208
sustained loading, 161–168
empirical approach, 166–168
rate process approach, 165–166
viscoelastic model approach, 164–165
Sverdrup, H.U., 91
SWSB, 14

T

Taiwan Island, **5**, 23, *24*, 25, 225–226, *230*
Taiwan Strait
geology, 23–28, *29*
Taiwan Strait Crossing, *see* TSC
Taiwan Strait Transportation Corridor, *see* TST
Tatsuoka, F., 162
territorial claims, 4–8
Thiers, R.G., 209–210
thermocline, 50
threshold gradient, 103
tides
diurnal, 41–42
origins, 40–41
semi-diurnal, 41–42
tidal currents, 42–43
Triassic, 12, 20
T-S diagram, 49
Tsai, C.F., 198
TSC, 225–226, *228*, *229*, 230, 232
construction types, 233–238, **234**, **235**, **237**, **238**
geotechnical parameters, estimation, 226, *227*
route selection, 226, *228*, 229–231
seismic resistance guidelines, **235–236**
TST, 225–238
tsunamis, 33, 61–64

U

UC, 95, 105–106, 209
under-consolidation, *see* UC
Unified Soil Classification System, *see* USCS
United Nations, **4**, 6
upwelling, 20, 48, 70
USCS, 79, 81, 119

V

van der Waals forces, 72, 74
Van Eekelen, H.A.M., 209
vertical effective stress, 96, 124, 207
vertical mixing, 48
viscoelasticity, 164
von Rad, U., 82
Vucetric, M., 189
Vyalov, S.S., 161

W

waves
deep water, 56–57, 61
development of, *53*, 60, **62**, 209
harmonic, 56

heights, 53, 59, 61
properties, *55*
shallow water, 57, 63
surface, 53–54, 56, 59, 206–208
theories, 54–61
wave dispersion equation, 56
wave loading, 174, *202*
wind, 60, **61**
wave-seabed interaction
Airy theory, 203–206
deformable seabed, 207–209
linear theory, 203–206
rigid seabed, 206–207
Wilson Cycle, 14
wind speeds, 44, 60, **61**, 64–65
WMO, 61, **62**
Woods, R.D., 197
World Meteorological Organization, *see* WMO
Wu, F.T., 24, 165, 197

Y

Yangtze Plate, 11–12, *13*
Yin, J.H., 96, 226